Privacy and Security Challenges in Cloud Computing

Cognitive Approaches in Cloud and Edge Computing

Series Editors:
Niranjanamurthy M and Ramiz Aliguliyev

Privacy and Security Challenges in Cloud Computing
A Holistic Approach
T. Ananth Kumar, T. S. Arun Samuel, R. Dinesh Jackson Samuel, and M. Niranjanamurthy

Ethics in Information Technology
A Practical Guide
Gajanan K. Awari and Sarvesh V. Warjurkar

For more information about this series, please visit: www.routledge.com/
Cognitive-Approaches-in-Cloud-and-Edge-Computing/book-series/CACAEC

Privacy and Security Challenges in Cloud Computing
A Holistic Approach

Edited by
T. Ananth Kumar
T. S. Arun Samuel
R. Dinesh Jackson Samuel
M. Niranjanamurthy

CRC Press
Taylor & Francis Group
Boca Raton London New York

CRC Press is an imprint of the
Taylor & Francis Group, an **informa** business

First edition published 2022
by CRC Press
6000 Broken Sound Parkway NW, Suite 300, Boca Raton, FL 33487-2742

and by CRC Press
4 Park Square, Milton Park, Abingdon, Oxon, OX14 4RN

© 2022 selection and editorial matter, T. Ananth Kumar, T. S. Arun Samuel, R. Dinesh Jackson Samuel and M. Niranjanamurthy; individual chapters, the contributors

CRC Press is an imprint of Taylor & Francis Group, LLC

ISBN: 978-1-032-11355-5 (hbk)
ISBN: 978-1-032-11426-2 (pbk)
ISBN: 978-1-003-21988-0 (ebk)

DOI: 10.1201/9781003219880

Typeset in Times
by codeMantra

Contents

Preface

Data security has been a significant issue in the IT field consistently. In the cloud computing environment, it becomes particularly severe because data are located in different locations even across the globe. This book discusses a comparative research analysis of existing research on data security and operational security agitations in cloud computing. The book provides readers with state-of-the-art cloud computing specifically for privacy and security issues. The book will include all the prerequisite methodologies so that new researchers and practitioners will find it very useful.

Chapter 1 deals with security-enhanced cloud techniques for serverless computing. The traditional cloud computing models are more complex in dealing with real-time data streams. The platform provides a place for all run-time applications to process. To improve the function of latency and efficiency of a system, new serverless computing is introduced. Serverless computing uses function-as-a-service (FaaS) models. FaaS liberates designers from the challenging work of working out or keeping up an unpredictable framework.

Chapter 2 reveals the web of cloud computing with threats and vulnerabilities. The migration from client/server models to service-based models means that the technical departments will have to revise their processes and procedures for implementing, developing, and distributing computer technology and applications. This chapter addresses an essential gap in our understanding of cloud computing by offering an overview of the cloud's security risks, threats, and vulnerabilities. Additionally, this chapter includes many elements such as threat actors, risks, vulnerabilities, and ramifications from real-world attacks and breaches.

Chapter 3 deals with the security and privacy provocation of data in cloud computing. Cloud computing is an on-demand resource sharing paradigm that carries sensitive information in cloud computing, covering a wide range of areas. These sensitive data on experiencing an attack can result in the crash of the entire system. This attracts the efforts of cybersecurity professionals to concentrate greatly on the security and privacy of sensitive data in cloud computing.

Chapter 4 addresses the networking security incitements in cloud computing. According to the community of requests to accept the library of common configurable registered attributes, these attributes can be quickly set and loosened through light operations or collaboration with professional cooperative vendors, as quoted by the National Institute of Standards and Technology. The convenience and cost adequacy of this model are its two fundamental benefits. While there are still concerns about security and vendor lock-in, this model has many advantages. This chapter delves into some of the fundamentals of cloud computing, including its forms, services, challenges, and security concerns.

Chapter 5 reports the major security concerns of virtualization in cloud computing. *Virtualization* is a technical innovation that distinguishes functions from the underlying hardware. Attackers may breach VM infrastructures and get access to different VMs on the same system. A *hypervisor* is a software program that exists between host virtual machines and hardware components. This chapter describes the cloud-based intrusion on the virtualized infrastructure.

Chapter 6 deals with operational security agitations in cloud computing. Being vulnerable to attacks, cloud computing deteriorates the growth of business and the economy, as most companies rely on cloud computing for their business processes and financial transactions. The increase in advanced technology for launching different types of attacks is a motivator in identifying technical solutions for improving the strength of the authentication process and safeguarding the cloud resources' data assets. This chapter explains the different types of vulnerability of cloud computing and various methodologies using cryptographic algorithms proposed to break these vulnerabilities and strengthen the security policy of cloud computing.

Chapter 7 describes the secure data storage and retrieval operations in cloud computing. This chapter proposes a new authentication system focusing on lightweight wireless devices and transferring information to the cloud processing without disclosing the collected data to the provider. This technique optimizes data storage and retrieval. It optimizes the computation and communication overhead, which in turn reduces the cloud service charges.

Chapter 8 takes a deep dive into popular security models and strategies of cloud computing. Cloud computing emerges as a horizon by serving critical applications. However, the unavailability of standard security standards and models makes cloud computing security more puzzling while transferring sensitive information across the cloud. This chapter aims to enlighten some of the novel security models and approaches in cloud computing.

Chapter 9 deals with quantum computing and quantum cryptography in cloud computing. The power of quantum computers can break the computations that are incredibly complex because of the superposition bits used in these computations. When it comes to the development of quantum computers, it is essential to remember that the integrity of encrypted data is in jeopardy right now. This chapter addresses quantum cryptography, which uses quantum mechanics to encrypt and transport data in a way that cannot be hacked.

Chapter 10 describes the extensive exploration of privacy and compliance considerations for hybrid cloud deployments. Public cloud and private cloud domains combine to form a platform known as hybrid cloud, which provides deployment methods to carry out various applications in different dimensions. The main issues faced by the hybrid cloud are minimal security controls and risk management. This hybrid storage service has opened up the scope for progress in organizational development. This chapter aims to highlight advanced privacy and security models for hybrid cloud models with state-of-the-art techniques as a functional discipline.

Chapter 11 addresses the threat detection and incident response in cloud security. Cloud security is critical for cloud storage providers because they must safeguard sensitive information while adhering to several regulatory obligations. Even though the data stay the same, the regulatory obligations change depending on where the data are stored. Incident response (IR) is the process of minimizing the impact of a discovered attack or compromise in a system or network.

Chapter 12 deals with the cyber artificial intelligence platform for cloud security. In the cloud, to ensure the security of the stored information, there are many models and platforms to protect the various essential data and the information. Artificial intelligence-based approaches will provide improved cyber defense capabilities in the cybersecurity environment and help opponents refine cyber threats. Cyber artificial intelligence is a significant change in our potential to safeguard sensitive data networks and digital environments. To authenticate several organizations' stored data, the cyber artificial intelligence framework is used for cloud security.

Chapter 13 proposes an enhanced hybrid and highly secure cryptosystem for mitigating security issues in cloud environments. In this chapter, a hybrid algorithm called enhanced hybrid privacy and secure (EHPS) algorithm is proposed, implemented, well evaluated, and compared with its peers. By understanding the positive features of hybrid algorithms, the proposed algorithm integrates the features of Advanced Encryption Standard (AES), Data Encryption Standard (DES), cipher block chaining (CBC), and Triple DES algorithm. Moreover, the reduced encryption and decryption times of the proposed algorithm compared with hybrid cryptosystem make the proposed algorithm a positive choice when a cloud environment has to be implemented with reduced encryption/decryption times and enhanced safety.

Editors

T. Ananth Kumar is an assistant professor at IFET College of Engineering, affiliated with Anna University, Chennai, India. He has published papers in various national and international conferences and journals. His fields of interest are networks on chips, computer architecture, and ASIC design. He is a life member of ISTE and some membership bodies. He has many patents in various domains, and has written many book chapters for Springer, IET Press, and Taylor & Francis Group.

T. S. Arun Samuel is currently an associate professor in National Engineering College, Kovilpatti, Tamil Nadu, India. He has authored more than 67 international journal articles, book chapters, and conference proceedings. He is a life member of Institute of Engineering (IE), India, and IAENG, and is a member of IEEE.

R. Dinesh Jackson Samuel is a postdoctoral researcher in the Faculty of Technology, Design and Environment in Oxford Brookes University, Oxford, UK. He worked as a teaching cum research assistant at Vellore Institute of Technology University. His research interests include medical imaging, computer vision, pattern recognition, video analytics, and machine learning. He has filed two patents with Intellectual Property India. He has authored/co-authored international journal articles, book chapters, and conference proceedings. He is on the reviewer board of many peer-reviewed journals such as IEEE Access, Springer, and Elsevier. His background in computer science, biomedical imaging, and machine learning contributes to his mindful and competitive approach.

M. Niranjanamurthy is currently working as an assistant professor in the Department of Computer Applications, M.S. Ramaiah Institute of Technology, Bangalore, India. He has more than 12 years of teaching experience and 2 years of industry experience as a software engineer. He has published 80 papers in national/international conferences/journals. He is working as a reviewer in 22 international journals. He is a member of IEEE, a life member of International Association of Engineers (IAENG), and a member of Computer Science Teachers Association (CSTA). His areas of interest are data science, ML, e-commerce and m-commerce related to industry internal tool enhancement, software testing, software engineering, web services, web technologies, cloud computing, big data analytics, and networking.

Contributors

Hamid Ali Abed AL-Asadi
Communications Engineering
  Department
Iraq University College
Basra, Iraq
and
Computer Science Department
University of Basrah
Basra, Iraq

A. Ashwitha
Department of ISE
M. S. Ramaiah Institute of Technology
Bangalore, India

Krishnan Ayyappan
Cloud Engineer
Amazon
Seattle, WA, US

P. Manju Bala
IFET College of Engineering
Villupuram, India

Jalel Ben-Othman
L2S Lab CentraleSupélec
Université de Paris Sud et Université de
  Paris 13
Rue Juliot Curie, France

W.T. Chembian
Department of Computer Science and
  Engineering
Gojan School of Business and
  Technology
Chennai, India

K. Dhanalakshmi
IFET College of Engineering
Villupuram, India

P. Divya
IFET College of Engineering
Villupuram, India

Amer S. Elameer
University of Information Technology
  and Communications
Baghdad, Iraq

G. Fathima
Adhiyamaan College of Engineering
Hosur, India

P. Hemalatha
Department of Computer Science and
  Engineering
IFET College of Engineering
Villupuram, India

S. Jayalakshmi
IFET College of Engineering
Villupuram, India

B. Jegajothi
Department of Information Technology
Sri Venkateswara College of
  Engineering
Chennai, India

S. Jeyapriyanga
Bharath Institute of Higher Education
  and Research
Chennai, India

L. Jerart Julus
Department of IT
National Engineering College
Kovilpatti, India

A. Ranjith Kumar
School of Computer Science and
Engineering
Lovely Professional University
Bangalore, India

M. Satheesh Kumar
Department of Electronics and
Communication Engineering
National Engineering College
Kovilpatti, India

K. Suresh Kumar
IFET College of Engineering
Villupuram, India

R. Senthil Kumaran
IFET College of Engineering
Villupuram, India

S. Maheswari
National Engineering College
Kovilpatti, India

Darshana A. Naik
Department of CSE
M. S. Ramaiah Institute of Technology
Bangalore, India

N. Nithiyanandam
Bharath Institute of Higher Education
and Research
Chennai, India

D. Prabakaran
IFET College of Engineering
Villupuram, India

K. Selva Banu Priya
Bharath Institute of Higher Education
and Research
Chennai, India

S.P. Priyadharshini
Bharath Institute of Higher Education
and Research
Chennai, India

Satheeshkumar Rajendran
Security Operations Centre
Rakuten Mobile
Setagaya City, Japan

R. Rajmohan
IFET College of Engineering
Villupuram, India

R. Ramani
PSR Engineering College
Sivakasi, India

A. Ramathilagam
PSR Engineering College
Sivakasi, India

S. Ramesh
Department of Computer Science and
Engineering
Krishnasamy College of Engineering &
Technology
Cuddalore, India

B. Sheik Mohamed
EY LLP
Chennai, India

A. Shenbagavalli
Department of Electronics and
Communication Engineering
National Engineering College
Kovilpatti, India

G. Shruthi
Department of ISE
M. S. Ramaiah Institute of Technology
Bangalore, India

N. Sivaranjani
Bharath Institute of Higher Education
and Research
Chennai, India

K.G. Srinivasagan
Department of Computer Science
Engineering
National Engineering College
Kovilpatti, India

Pramod Sunagar
Department of CSE
M. S. Ramaiah Institute of Technology
Bangalore, India

S. Usharani
IFET College of Engineering
Villupuram, India

A. Valarmathi
Department of Electronics and
Communication Engineering
University VOC College of Engineering
Thoothukudi, India

R. Velvizhi
Bharath Institute of Higher Education
and Research
Chennai, India

K. Venkatesh
Vel Tech Rangarajan Dr. Sagunthala
R&D Institute of Science and
Technology
Chennai, India

1 Security-Enhanced Cloud for Serverless Computing and Its Applications

L. Jerart Julus
National Engineering College

Krishnan Ayyappan
Amazon

CONTENTS

DOI: 10.1201/9781003219880-1

1.1 INTRODUCTION

Computing refers to the activity of building and benefiting from computing infrastructure. This includes the development of software, hardware, and the research of algorithms. In the last 10–15 years, a new model of computing, called "cloud computing," has increased in usage and popularity [1]. Cloud is nothing but the collection of computing resources (such as powerful computers, servers, and so on) located in a particular place, which we call data centers. Unlike the personal computing environment, these data centers are owned by cloud service providers and not by the users themselves.

Cloud computing is the process of delivering the services (such as databases, storage, computing capacity, and so on) needed for software applications through the Internet [2]. The categories of cloud are public cloud, private cloud, hybrid cloud, and virtual private cloud. When cloud services are accessible by the public for their usage, it is called public cloud. For example, Google search engine, Gmail, Google Drive, and related services are hosted on the cloud and are available to the public. When cloud resources are set up for the use of a particular organization alone, then it is called private cloud. For example, a web application hosted on a university's server(s) to enable the students to see their assignments and scores. Hybrid cloud is a name given to the service which has the arrangement of public and private cloud [3]. That is, a set of non-critical applications/data will be hosted in public cloud. And the critical applications/data will be hosted in private cloud. In virtual private cloud, resources will be provisioned from a public cloud; however, access will be restricted only to the employees of an organization. So, virtual private cloud will be present inside the private cloud [4]. For example, organizations host their applications in AWS Cloud, Google Cloud Platform, Microsoft Azure, etc., but have restricted access to their employees alone.

Until the early 2000s, businesses had to spend a lot of money not only on buying expensive hardware but also on maintaining them. The introduction of cloud computing in 2006 was a breakthrough in the IT industry. Since then, cloud computing has revolutionized the entire process involved in setting up IT infrastructure for businesses [5].

Serverless computing enabled businesses to utilize the cloud to its fullest potential, thus taking server maintenance out of the picture. This helped in focusing on the product than spending time on setting up the infrastructure. Usage of multi-tenancy architecture helped optimize the costs involved in using the cloud [6]. (Multi-occupancy is an engineering design wherein a solitary occasion/piece of programming can have a large number of tenants. This is analogous to virtual hosting.) Cloud providers charge the customers only for what they use. This pricing model is very dynamic. For example, customers using the cloud for storage are charged for every gigabyte of data.

With the increasing usage (and demand) of cloud computing, companies such as Amazon, Google, Microsoft, Oracle, and so on brought up a business model based on it. That is, these companies started offering cloud computing resources and charge customers for using their infrastructure. As the basic setup can be completed within minutes, the services provided by these companies gained popularity [7]. Now, a lot of businesses/companies around the world use cloud computing to a greater extent due to its plug-and-play nature and its flexibility of accessing from anywhere. This has helped to continue operations during the pandemic situation, even when a majority (or all) of the workforce are working from home.

1.2 LITERATURE SURVEY

Cloud computing is the process of delivering the services (such as databases, storage, computing capacity, and so on) needed for software applications through the Internet [8]. In cloud computing, the user/client, and the hardware on which the application is run are decoupled and can be accessed/utilized remotely. Usage of these resources is available on demand, and there are no upfront costs. You can pay for what you use (pay-as-you-go model) [9]. Upgrading or downgrading the cloud computing resources can be done with a few clicks. However, changing the hardware is almost impossible in server-based computing due to the costs involved in it. Traditionally, in server-based computing, applications will be run on a server (and not by the client machine) that is hosted on premises. User/client must connect to the server through intranet/Internet (mostly, intranet) to access the application/storage/compute capacity. Servers are very expensive than PCs/workstations. In addition to the hardware, you will have to spend on the server software too. Despite all these expenditures, servers become slow and incapable of running resource-hungry applications and do not last longer than 5–6 years, on average [10]. First, during the 5–6 years of usage, periodic checks, updates, and monitoring are a cumbersome task that requires IT expertise and time. In case of issues with the servers, the cost of fixing/replacing them is very high. Second, a single point of failure is created, as the same (or same set of) server(s) is used for storage, database, and compute capacity. The impact of server failure or data loss will be huge. Of course, creating manual backups or scheduling automatic backups is a solution. But this solution comes with additional cost, which adds up to the financial burden. Third, when all the application(s) are hosted in the on-premises server, troubleshooting the issues in the server requires manpower. That is, you will need to set up a dedicated IT team to take care of the troubleshooting and maintenance tasks [11]. Finally, the process of transitioning to a server-based network or to a serverless network from a server-based network is expensive and complex. Additionally, performing this migration without any downtime to the running applications is very difficult. Using serverless computing helps you overcome all these disadvantages/difficulties in running the application in a server-based environment. The major types of cloud computing are infrastructure as a service (IaaS), platform as a service (PaaS), software as a service (SaaS), and serverless. Using serverless computing, instead of running applications in their servers/workstations (server-based computing), they can be run on servers that are present in data centers of cloud service providers such as Amazon Web Services (AWS), Google Cloud Platform (GCP), Microsoft Azure, and so on. The term "serverless" is a misnomer, as the applications will still run on a server. But the only difference is that the user does not own, rent, lease, or purchase the server.

By choosing a serverless approach over other conventional approaches, you not only save a lot of money but also have the advantage of elasticity and increased productivity. (Elasticity is the ability of serverless systems to scale themselves in terms of processing power based on the needs [12,13]. For example, additional web server instances are created when the number of users accessing the website is more.)

1.3 ARCHITECTURE OF SERVERLESS COMPUTING

Serverless architectures try to improve the old-style customer worker structure by lessening handling load and permitting effective scaling. In general, the older method follows the client–server structure. This implies that a server is facilitated on a PC to convey the information with the clients and the rights of sharing are owned by the server itself. The important factors that have to be considered by the servers are load balancing and scaling [14]. The new method of serverless structure enhances scaling and handles load very efficiently.

The best example that can be provided for a server is of a food restaurant. The server who serves the food is capable of providing all the dishes ordered by the clients. The same method is followed by the server which has the capability to serve all the signals that are provided to it. The rights to serve and control over the signal are monitored by it. When the number of clients who visit the restaurant increases, the wait time increases, and the challenge to serve all the clients within a shorter period arises. Since the entire client population approach at the same time, there is a possibility of server overloading [15]. On the off chance that a server crashes under any circumstances, your whole site will crash and will become difficult for clients and service providers to access it. Numerous organizations commonly have their creation database put away on a similar server as their site, implying that if the server crashes you will likewise be not able to interface with their database. Serverless structures are worked by eliminating the information base from the server and running individual administrations and capacities on interest. Figure 1.1 shows different vendors using the serverless architecture for future technological enhancement. The vendors

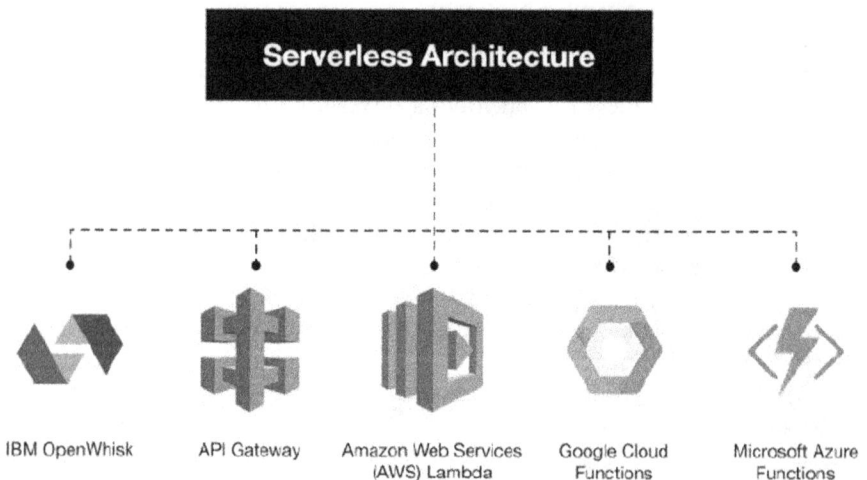

FIGURE 1.1 General serverless architecture supported by many vendors.

use these features for their services [16]. The platform used by all these vendors is function as a service (FaaS). The money is spent for utilizing the service based on the customers' execution time over these functions and for memory allocation.

The services are categorized under two management schemes as shown in Figure 1.2. The first method uses file-based management, and the files such as documents in text, multimedia, and other supporting formats are managed here. The second method is asset management. Asset management is the route toward making, working, keeping up, and selling resources. Asset management demonstrates an exact method to manage the organization and affirmation of huge worth from the things that a social affair or substance is liable for, over their whole life cycle [17]. All the governance is converted into a digital format which requires a greater amount of storage space and management capabilities.

Figure 1.3 shows the different digital asset platforms which are used for text, video, audio, and image. Coordinated administrations for ingestion, explanation, inventorying, capacity, recovery, and circulation of digital assets are done through

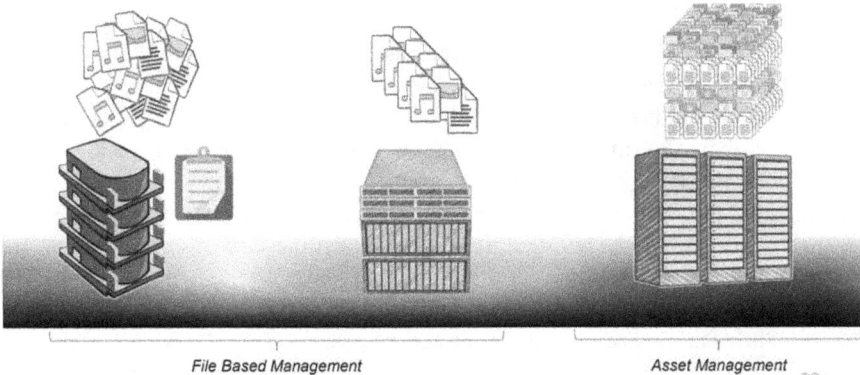

FIGURE 1.2 Different management requirements of users and customers.

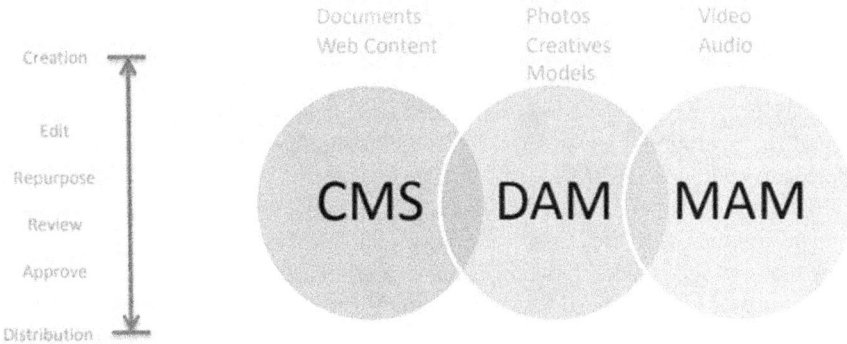

FIGURE 1.3 Different digital asset platforms.

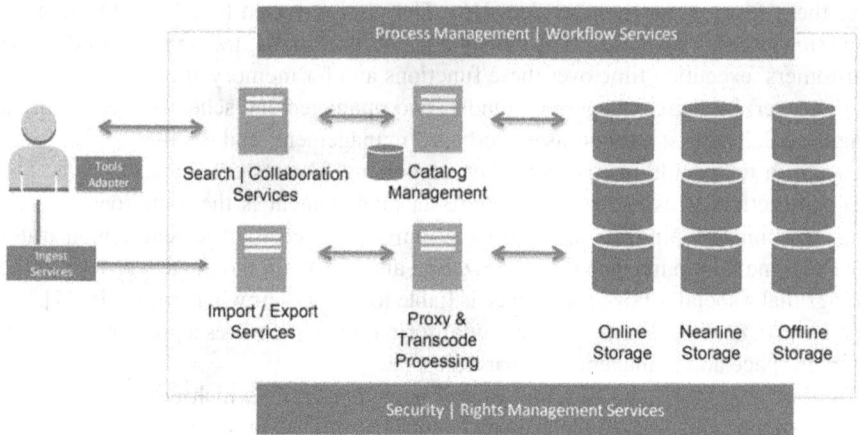

FIGURE 1.4 Architecture of a general digital asset management system.

Certified Wealth Strategist (CMS), digital asset management (DAM), and media asset management (MAM) [18].

The general architecture of a DAM system is illustrated in Figure 1.4. The architecture fills with two different layers. The process management holds and controls the different workflow services. The bottom layer is the security management that provides rights for utilizing the web-based services. Three different storage premises are used in the system: the online storage system, nearline storage system, and offline storage. These storage systems are directly connected with the ingest services. The tool adapter helps to connect the client with the storage system through catalog management [19]. The security management will be processing the proxy and transcode for every process. The process is mainly dealing with import and export services. The encrypted data can be stored in any one of the storage based on the needs of the application, and the data are processed securely.

1.3.1 DIGITAL ASSET MANAGEMENT

A DAM framework utilizes a setup of utilizations connected to give a consistent encounter to the client. It is an assortment of a huge component that has conscientiously been chosen to address the issues and spending plans of the venture. To expand the operational advantages, it is worthwhile for these applications to trade information with the DAM framework [20].

To expand the operational advantages, it is worthwhile for these applications to trade information with the DAM framework. The DAM is not a single system; it has many features and is a combination of many networks. Figure 1.5 shows the general architecture of DAM. The DAM has different architectures, and the architectures are differentiated based on the database model, business, and presentation model. One of the essential elements of any DAM framework is to ease the clients' needs to perform dull undertakings. The success of the system is based on the interface

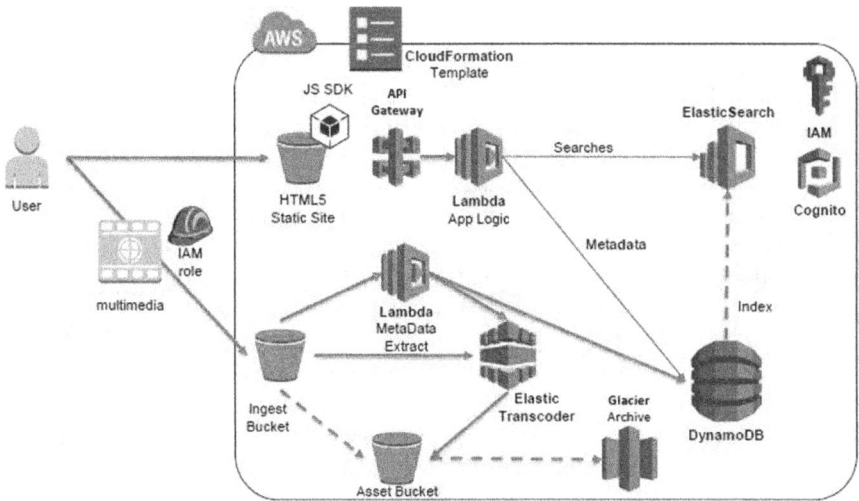

FIGURE 1.5 Simple architecture of DAM.

between the users. In case an unreasonable number of changes are made to the working techniques, customers can feel antagonized by asset management. Confined scale primers are significant to examine work measure issues. For the customers, the goal is to feel invigorated by the odds presented by the system, not to hate what could be viewed as a structure compelled to decrease observing levels [21]. For the clients, the objective is to feel energized by the changes introduced by the framework, not to detest what could be seen as a framework forced to diminish monitoring levels.

1.4 CENTRALIZED SERVERS TO MULTI-LAYER

Recalling through the authentic background of large business figuring, everything started with the concentrated PC and fundamental clients reliant on the video terminal or visual showcase unit (VDU). More modest centralized servers (the minicomputer) were created to suit the requirements of medium-sized organizations, and afterward, IBM dispatched the PC. It allowed even the most diminutive business to make capable-looking correspondence and to run accounts programming. As the costs of the work territory PC dropped to insignificant more than the work territories they sat on, business customers mentioned a comparable UI as the PC [22]. A genuine model is the word processor. Simultaneously, it empowered the client to be more inventive with their design. This capacity to offer additional opportunities stretches out to asset management. The centralized server management system mostly deals with the complex structure of asset management. The DAM has two elements: file processing and creative processing. There is a lot of data processing, identifying with metadata inquiries and updates. Three layers are present in the structure: the presentation layer, business layer, and data layer. The layer has no incredible issue to work and keep up because there is only a web worker. The design of the DAM can be split into a different pack known as business rationale [23]. The standards designed for controlling the information, and the associations with outsider applications, would all be able to dwell on a different server application that runs the center asset management software.

1.4.1 SEARCH

The search is an important feature of asset management. The customer can use the web search tool to find content inside the archive. Searching is different compared to the normal text search and natural language processing.

1.4.2 PROCESS MANAGEMENT

The system gives a phase to robotize the business measures. It ought to have essential highlights such as e-mail alarms. It might incorporate more complete facilities.

1.4.3 JOB AND DESIGN MANAGEMENT

The source management application should offer various administrations to help increase the capacity of the executives. These will incorporate capacity organizations; various leveled stockpiling the board to deal with the circle, close line and disconnected, and media the executives for removable and simple stockpiling.

1.4.4 CLIENT MANAGEMENT

This module deals with the clients of the framework, with offices to customize the web interface, partner with projects, and change individual arrangement settings.

1.4.5 STORAGE MANAGEMENT

Much substance may as of now exist in an advanced structure. Records can be ingested by FTP or other documents move from customers or accomplices. Outside the center resource there are various satellites for the board application. There will be inheritance situations, similar to the rights of the board and the back-office frameworks [24].

Google Cloud Functions is a serverless execution environment for building and connecting cloud services. With the help of cloud functions, we can simply write the backend processing code as APIs and invoke the functions whenever we need it. There is no backend server setup needed here. This makes it a serverless environment and easy to set up so that we can concentrate more on code rather than setting up a whole environment.

1.4.6 GOOGLE CLOUD FUNCTIONS

Another main reason to use cloud functions is to execute the backend code in a secure and trusted environment so that the application will be safe from attacks.

As the cloud function is one of the parts of GCP, we can use other Google Cloud Services from our cloud functions itself, as the services can only be used from a trusted environment.

Google Cloud Functions can be written using
- JavaScript
- Python 3
- Go
- Java.

In addition to processing the incoming requests, cloud functions can also be used to do any task based on some cloud events that trigger the cloud functions. As the cloud functions can access other Google Cloud Services, we can set triggers for some events on different Google Services and execute some functions based on those triggers such as greeting a new user in our application, and deleting all data associated with a user when the user deletes their account.

For example, let us consider that we have a social media application such as Facebook or Instagram where the users can share media content such as images, videos, and so on. Figure 1.6 shows the Google Cloud Feature Trigger for file upload to Bucket Storage. We've to make sure that no offensive image or media is posted on the site as per our regulations. So, in order to find and neglect the offensive media, we can set a Cloud Trigger Function on our cloud storage that whenever a media file is uploaded in the storage, it triggers a cloud function which will take the file and check for any offensive content in that image with the help of any image processing services such as Google Cloud Vision and take necessary actions. If the media file contains no offensive content, it will be posted or else either the file post gets removed or the media will be blurred with the help of other services such as ImageMagick depending on our terms and conditions and posted on the platform.

1.5 FEATURES PRESENT IN CLOUD

The feature of the cloud is based on the serverless computing technologies to be included in the analysis. The market potential is mainly targeted on the security of the data storage on the cloud and also on having a good indigenous server in their own locations. The explanations for the market needs are discussed briefly below.

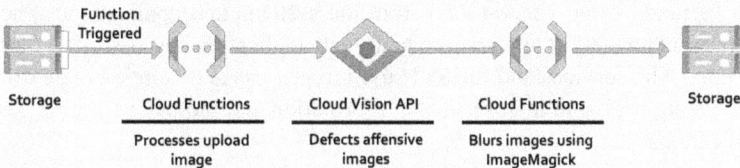

Storage	Cloud Functions	Cloud Vision API	Cloud Functions	Storage
	Processes upload image	Defects affensive images	Blurs images using ImageMagick	

FIGURE 1.6   Google Cloud Function Trigger on file upload to Bucket Storage.

1.5.1 SCALABILITY AND ELASTICITY

There are two normal scenarios in which the capability of cloud computing can be tested for research. The user-related scenario is tested first. A dataset that is created by Sasha is too big or has to be mixed several times, which makes it impractical to work in your own data center. In such situations, it would be a better option to make the dataset accessible in a backend system. The data can be replicated in many backend servers around the world. The data center connects all backend servers for their process, where the task operation is unknown to the client. The second scenario is the bandwidth required to transmit data to several users. These scenarios involve the need to conduct an unexpected task in a very short period of time because part of your analysis has unexpectedly become important to media or policy problems or because it is needed by your client, founder, or stakeholder [25].

1.5.2 PROVISION FOR OWN SERVICE

The client has an option of choosing their required service from the available resources. The service aims to provide a simple management tool to reduce complexity in the process. Users need to have a clearer understanding of the choices and relevant solutions as the freedom of choice increases; thus, many forums have provided different online specialized trainings and accreditations to enhance the level of knowledge.

1.5.3 INTERFACES FOR APPLICATION PROGRAMMING

The communication between the software layers and the infrastructure based on the application is known as an application programming interface (API). The API provides an opportunity to communicate with two different types of software applications. Many real-time applications are integrated by API. These API advantages are used in serverless computing to provide an interface between hardware and cloud uploads. The Cloud API lays a platform to access different domain-based data and interface to a cloud environment. The process of each data is done indigenously, and the integration with different applications is done through API.

1.5.4 MONITORING AND ASSESSING OUTPUT

Cloud offers observability, to contact data and entries that allow empirical observations to be made using general data from the systems and applications. There are many different formats of applications, and each application has its style in representing the data. The services and the tools used to process and store data are different. Many tools are available to observe the information and modify its access based on the applications.

Cloud-based serverless computing requires a high-speed data connection with users, so optical communication using FBMC for 5G connectivity service helps achieve this. Few methods that are used so far are OFDM modulation and phase-shift keying methods to practice serverless computing [26].

1.6   PROVIDING GREEN IT

Serverless computing facilitates green IT to save the environment by sharing the available resources and increasing the quality of service.

1.6.1   SHARING OF RESOURCES

Green IT helps the environment to maintain sustainability from IT sectors. It uses non-hazardous materials during the manufacture of components in computing devices. To minimize the amount of power consumption and increasing the life cycle of the hardware and services. The modern innovation in IT mainly focuses on utilizing resources with less cost and good profit. The virtual access of devices provides an efficient way of using remote systems. Serverless computing is a step up for a green environment. In general, server computing consumes 18% of energy and produces nearly 640 million tons of carbon emission every year. Serverless computing provides less utilization of power, due to the finer infrastructure and cooling system.

1.6.2   THE CODE QUALITY

Event-driven computing is the main idea behind serverless computing. Every event relates to various functional executions. Large numbers of developers are encouraging serverless computing, which uses backend services. The complexity of the system is reduced because it uses granular functions. It supports multi-tasking to test different versions of code.

1.7   APPLICATIONS OF CLOUD COMPUTING

While the idea is relatively recent, the literature has suggested several real-world serverless applications.

1.7.1   COLLABORATION AND ANALYTICS IN REAL TIME

Serverless services are stateless in nature and are more useful for storing and accessing real-time data which support autonomous vehicles. Conversation with chatbots for the support chain help desk is one of the future applications. A serverless XMPP-based instant messaging approach was also implemented in [27]. The remote location invention such as different space missions depends on the serverless service to monitor and send information to long distances in space. Military operations of many countries and their information are kept secured using different indigenous storage devices to enhance security. The service collects, analyzes, and then reflects the analytics of data. Serverless computing's auto-scaling feature makes it possible to manage concurrent large data streams.

1.7.2   SYSTEMS OF URBAN AND INDUSTRIAL MANAGEMENT

The pay-as-you-go serverless service model paved the way for different budget-restricted urban and industrial management systems to be adopted and implemented.

Serverless systems have also been used for applications for urban disaster recovery. The authors of [28] proposed a group formation approach after disasters using services without servers.

1.7.3 Science Informatics

Serverless computing has not been explored as an appealing solution for scientific computing applications, although several studies have based their attention on serverless services for those applications. We assume the disagreement lies in the belief that there is a wide variety of scientific computing and its implementations and there are some fields of this area in which it is possible to use serverless services.

1.7.4 Learning Computer

In particular, machine learning and neural network-based approaches are currently one of the most enticing research trends. There is a range of problems that need to be tackled, as the subject of serverless computing is relatively new. Some of these issues are not peculiar to the model of serverless computing and can occur in any computer environment, such as security and privacy, caching, or pricing. Some types of problems are also unique to serverless services, such as feature invocation or intercommunication functions. It is especially important to address the complexities of the extensive adoption of serverless services, as there has been a debate between different experts about the advantages of migration from server approaches to server services and non-servers. The disadvantages and weak emulation outcomes found are all rooted in current problems that are specific to serverless services and have to be discussed in future research.

1.7.5 Intra-Communications and Exchange

The structure of a few capacities that cooperate to give the ideal desired functionalities is serverless software. To do this, the jobs need to interface with each other and then somehow share their information or the state. In other cloud suppliers, this data sharing framework is refined by network executives. In IaaS, each virtual machine may use network addresses to send and receive data via point-to-point networking. In today's serverless infrastructures, function-level network addressing is not possible, so two functions communicate through intermediary storage systems. In contrast to that of direct network links, this burdens a large amount of load on comparatively sluggish and costly storage systems. Serverless service functions have some attributes that need to be addressed to implement an addressing scheme for them. Because of the auto-scaling of serverless computing, there could be many running invocations of the same function within different functions at any given time. Sometimes, the roles are short-lived. The short life of the functions implies that every addressing method should be quick enough to control the rapidity of the functions.

The key aim of the life of serverless services is its ability to optimize itself by providing copies of the functions of customers and assigning those copies to each order. There have been no predefined approaches as to where to physically position the copies of functions. Alternatively, inside the service provider, load balancing mechanisms decide on the current usage of computing nodes and the available resources. Computer programs use several cache levels to escape the bottlenecks and latencies of persistent storage. When a feature requests a user's data from the database, the platform typically caches the data to accommodate any new requests and to reduce the amount of expensive access to the database. It fits well with server services. In a serverless operation, however, the next process execution can be assigned to another accessible computing node that renders the function. The operation of management within serverless providers is a dynamic and resource-demanding task. For each of the infrastructures within the provider, multiple monitoring and provisioning operations are involved. In any computing service, whether serverless or not, security is an important concern.

1.7.6 PRIVACY AND DEFENSE

Different security problems are prevalent among serverless people and other facilities for the cloud. How to authenticate applications so that only legitimate ones can use the available functions is the key security problem in any serverless scheme. Amazon's Lambda is currently introducing such a framework that uses a bearer token authentication technique to establish the identity of the caller. The use of authentication tokens in the request header is a standard technique to counter these attacks. If the application is sent via an unsecured system, however, the attacker may simply delete the token from another program and reuse it. Services without servers are also vulnerable to replay attacks. The attacker is unaware of the content of the message being sent in these forms of attacks, but they are interested in the impact of transmitting the message.

The intruder then catches the request for safe function execution and replays it to sabotage the system's normal service. The infinite replay of a logout request to prevent users from accessing their desired service is an example of such an attack. A challenge-response protocol for serverless services must be implemented to remedy these types of attacks. There is also the question of permission for roles. This is fundamentally distinct from the authentication at the application level we discussed previously. The lack of a proper authorization scheme may pose serious threats to security applications. Recall that the ability to buy off-the-shelf functions is one of the benefits of serverless services, as we listed. Functions may be used without the approval of the program creator, without authorization schemes. To manage function-level authorization, a proper mechanism should be implemented. It is also possible to create a resource exhaustion attack against the provider itself. The future depends on serverless computing models to maintain good energy efficiency and fewer hardware requirements. This platform provides an opportunity for interconnecting many devices and maintains a green computing environment. The system provides good authentication and secures the file system from attackers.

1.8   CONCLUSIONS

Cloud computing has become an integral part of an organization's IT application architecture. A survey taken by a private organization shows that 81% of the organizations that have 1,000 employees or more use multiple platforms (such as on-premises, cloud, and so on) for hosting their applications and data. Organizations that are still using the on-premises application on a larger proportion are migrating to the cloud. Despite the challenge of migrating without (or minimal) downtime, the migration activities are being carried out at a steady pace. Increased usage of the cloud is saving millions for organizations. Additionally, maintenance is becoming easier in the cloud as the cloud providers release new features to automate maintenance tasks.

According to analysts, cloud computing will evolve and will incorporate things such as edge computing, the Internet of things (IoT), and so on. According to the prediction of another private entity, by 2025, almost 75% of the data generated will be from the cloud. Also, cloud usage of around 40% of the organizations will be edge computing, and these edge devices will be capable of running AI algorithms. Thanks to the cloud that it has helped employees to work from home and the students to learn from home. In other words, the cloud has enabled every organization (private sector and public sector) and educational institution to continue their operations without issues. As the cloud has given this flexibility to switch to a "work from anywhere" approach, it has helped organizations and institutions to focus on their performance without interruption. The approach taken during this pandemic situation to keep the businesses running has set new expectations in terms of flexibility, mobility, agility, and security. Even though these are not achieved immediately, these will be perspectives from which each cloud service's ability will be measured. Experts say that we are in the starting phase of a golden era of computing. In the future, a lot of AI and IoT technologies will be available in everyday life of ours to an extent that we don't feel that there is an extra device or technology around us (just like how we treat mobile phone signals or electricity these days). Just like the clouds in the sky, there is no limit for cloud computing in the future. Every reader of this book will be building a lot of applications in the cloud. Happy Cloud Computing!

REFERENCES

1. Mishra, Bhabani Shankar Prasad, Himansu Das, Satchidananda Dehuri, and Alok Kumar Jagadev (Eds.) *Cloud Computing for Optimization: Foundations, Applications, and Challenges.* Cham: Springer International Publishing (2018).
2. Lee, Kwangsu. "Comments on "secure data sharing in cloud computing using revocable-storage identity-based encryption"". *IEEE Transactions on Cloud Computing* 8, no. 4 (2020): 1299–1300.
3. Christidis, Angelos, Sotiris Moschoyiannis, Ching-Hsien Hsu, and Roy Davies. "Enabling serverless deployment of large-scale AI workloads." *IEEE Access* 8 (2020): 70150–70161.
4. Sarkar, Suvajit, Rajeev Wankar, Satish Narayana Srirama, and Nagender Kumar Suryadevara. "Serverless management of sensing systems for fog computing framework." *IEEE Sensors Journal* 20, no. 3 (2019): 1564–1572.
5. Lin, Changyuan, and Hamzeh Khazaei. "Modeling and optimization of performance and cost of serverless applications." *IEEE Transactions on Parallel and Distributed Systems* 32, no. 3 (2020): 615–632.

6. Baldini, Ioana, Paul Castro, Kerry Chang, Perry Cheng, Stephen Fink, Vatche Ishakian, Nick Mitchell, et al. "Serverless computing: current trends and open problems." In: Sanjay Chaudhary, Gaurav Somani, Rajkumar Buyya (Eds.) *Research Advances in Cloud Computing*, pp. 1–20. Singapore: Springer (2017).

7. Lynn, Theo, Pierangelo Rosati, Arnaud Lejeune, and Vincent Emeakaroha. "A preliminary review of enterprise serverless cloud computing (function-as-a-service) platforms." *In 2017 IEEE International Conference on Cloud Computing Technology and Science (CloudCom)*, pp. 162–169, IEEE (2017).

8. Rajan, R. Arokia Paul. "Serverless architecture-a revolution in cloud computing." *In 2018 Tenth International Conference on Advanced Computing (ICoAC)*, pp. 88–93, IEEE (2018).

9. Al-Sayyed, Rizik MH, Wadi'A. Hijawi, Anwar M. Bashiti, Ibrahim AlJarah, Nadim Obeid, and Omar Y. Adwan. "An investigation of Microsoft Azure and amazon web services from users' perspectives." *International Journal of Emerging Technologies in Learning* 14, no. 10 (2019).

10. Castro, Paul, Vatche Ishakian, Vinod Muthusamy, and Aleksander Slominski. "Serverless programming (function as a service)." *In 2017 IEEE 37th International Conference on Distributed Computing Systems (ICDCS)*, pp. 2658–2659, IEEE (2017).

11. Sewak, Mohit, and Sachchidanand Singh. "Winning in the era of serverless computing and function as a service." *In 2018 3rd International Conference for Convergence in Technology (I2CT)*, pp. 1–5, IEEE (2018).

12. Yussupov, Vladimir, Uwe Breitenbücher, Frank Leymann, and Christian Müller. "Facing the unplanned migration of serverless applications: a study on portability problems, solutions, and dead ends." *In Proceedings of the 12th IEEE/ACM International Conference on Utility and Cloud Computing*, pp. 273–283 (2019).

13. Lin, Changyuan, and Hamzeh Khazaei. "Modeling and optimization of performance and cost of serverless applications." *IEEE Transactions on Parallel and Distributed Systems* 32, no. 3 (2020): 615–632.

14. Kim, Young Ki, M. Reza Hoseiny Farahabady, Young Choon Lee, and Albert Y. Zomaya. "Automated fine-grained cpu cap control in serverless computing platform." *IEEE Transactions on Parallel and Distributed Systems* 31, no. 10 (2020): 2289–2301.

15. Sbarski, Peter, and Sam Kroonenburg. *Serverless Architectures on AWS: With Examples Using AWS Lambda*. Shelter Island: Manning Publications Company (2017).

16. Suresh Kumar, K., A. S. Radha Mani, S. Sundaresan, and T. Ananth Kumar. "Modeling of VANET for future generation transportation system through edge/fog/cloud computing powered by 6G." In: Gurinder Singh, Vishal Jain, Jyotir Moy Chatterjee, Loveleen Gaur (Eds.) *Cloud and IoT-Based Vehicular Ad Hoc Networks,* pp. 105–124. Hoboken, NJ: John Wiley & Sons (2021).

17. Fernandez, Obie. "Serverless: patterns of modern application design using microservices (Amazon web services edition)." preparation, https://leanpub. com/serverless (2016).

18. Gu, X., L. Yang, Y. Liang and S. Cao. "Design and implementation of digital assets management system based on WeChat," *In 2017 IEEE 2nd Information Technology, Networking, Electronic and Automation Control Conference (ITNEC)*, Chengdu, pp. 394–398 (2017). doi: 10.1109/ITNEC.2017.8284759.

19. Shailesh Kumar Shivakumar, "Digital asset management and document management," *In Enterprise Content and Search Management for Building Digital Platforms*, pp. 253–271, IEEE (2017). doi: 10.1002/9781119206842.ch8.

20. Rajan, R. Arokia Paul. "Serverless architecture-a revolution in cloud computing." *In 2018 Tenth International Conference on Advanced Computing (ICoAC)*, pp. 88–93, IEEE (2018).

21. McGrath, Garrett, and Paul R. Brenner. "Serverless computing: Design, implementation, and performance." *In 2017 IEEE 37th International Conference on Distributed Computing Systems Workshops (ICDCSW)*, pp. 405–410, IEEE (2017).

22. Carreira, Joao, Pedro Fonseca, Alexey Tumanov, Andrew Zhang, and Randy Katz. "Cirrus: A serverless framework for end-to-end ml workflows." *In Proceedings of the ACM Symposium on Cloud Computing*, pp. 13–24 (2019).

23. Jonas, Eric, Qifan Pu, Shivaram Venkataraman, Ion Stoica, and Benjamin Recht. "Occupy the cloud: Distributed computing for the 99%." *In Proceedings of the 2017 Symposium on Cloud Computing,* pp. 445–451 (2017).

24. Bharti, Urmil, Deepali Bajaj, Anita Goel, and S. C. Gupta. "Sequential workflow in production serverless FaaS orchestration platform." *In Proceedings of International Conference on Intelligent Computing, Information and Control Systems*, Springer, Singapore, pp. 681–693 (2021).

25. Kumanov, Dimitar, Ling-Hong Hung, Wes Lloyd, and Ka Yee Yeung. "Serverless computing provides on-demand high performance computing for biomedical research." arXiv preprint arXiv:1807.11659 (2018).

26. Julus, L. Jerart, D. Manimegalai, and S. Sibi Chakkaravarthy. "FBMC-based dispersion compensation using artificial neural network equalization for long reach-passive optical network." *International Journal of Wavelets, Multiresolution and Information Processing* 18, no. 01 (2020): 1941011.

27. Al-Masri, Eyhab, Ibrahim Diabate, Richa Jain, Ming Hoi Lam Lam, and Swetha Reddy Nathala. "A serverless IoT architecture for smart waste management systems." *In 2018 IEEE International Conference on Industrial Internet (ICII)*, pp. 179–180, IEEE (2018).

28. Hussain, Razin Farhan, Mohsen Amini Salehi, and Omid Semiari. "Serverless edge computing for green oil and gas industry." *In 2019 IEEE Green Technologies Conference (GreenTech)*, pp. 1–4, IEEE (2019).

2 Revealing the Web of Cloud Computing
Threats and Vulnerabilities

M. Satheesh Kumar
National Engineering College

Jalel Ben-Othman
Université de Paris

B. Sheik Mohamed
EY LLP

A. Shenbagavalli
National Engineering College

CONTENTS

DOI: 10.1201/9781003219880-2

2.1   INTRODUCTION

The concept of cloud in cloud computing is a metaphor for the Internet, similar to how real clouds collect water particles. It is a highly advanced data centre that provides data storage and security, and builds collaboration between employees. While decreasing the cost, it leads to making better decisions and changes the progress of small-scale businesses to large entrepreneurs [1]. The workload is managed by the networks that form the cloud, which is why the workload on the host machine is not too high while running an application. As a result, the user's demand for software and hardware remains reduced [2]. To make use of cloud storage, all we need is a web browser such as Chrome/Firefox. There are many aspects of cloud computing characteristics that make it one of the fastest growing markets [3].

2.1.1 CHARACTERISTICS OF CLOUD COMPUTING

Enterprises would be able to exploit the future gains as cloud computing technologies evolve both economically and technologically. Cloud computing has five distinct characteristics, which are described below.

2.1.1.1 On-Demand Self-Service

This service can be used as required and paid per usage. Think of it more like electricity. In essence, the cloud consists of utility computing [4]. We create an account or select our supplier, and can have access to their services any time. At the end of the month, we will be charged only for what has been used. This method of data collection and access gives us complete control over the utilisation and expenditure of the resources.

2.1.1.2 Broad Network Access

Using any computer with Internet connection, the user can access it from across the network. The cloud data can be available via web browsers anywhere we are, as well as on a desktop or handheld computer [5].

2.1.1.3 Resource Pooling and Multi-Tenancy

Cloud storage processes are designed to accommodate a multi-tenant environment. Multi-tenancy allows many users to share the same application or physical infrastructure while maintaining data security [6]. It is equivalent to residents in a residential complex who share the same house structure, but within the infrastructure, they each have their own apartment and security. When it comes to resource pooling, multiple users can share the same space, and services can be allocated, re-assigned, and dispersed as required. The user can be anywhere in the world and will get equal access as everybody else.

2.1.1.4 Rapid Elasticity

The cloud can broaden and shrink as much as possible without compromising either its users or their data [7]. If the company is seeing a surge in need, for example, the cloud will expand to meet the new requirements.

2.1.1.5 Measured Service

It can be examined how much the cloud is used by users. Most cloud service providers use a pay-as-you-go approach to ensure that their customers are billed for just what and how much they use, no more and no less [8]. Similar to what we pay for our electricity, water, and gas services, we are billed for the service utilized.

The goal of this chapter is to provide a comprehensive and complete overview of threats and vulnerabilities in cloud computing and to offer a comprehensive analysis of how to reveal them using case studies. The remainder of this chapter is structured as follows: Section 2.2 describes cloud deployment models. In Section 2.3, we provide a detailed illustration of cloud service models. In Section 2.4, we describe the specific threats in cloud environment. In Section 2.5, we explain the vulnerabilities in cloud. In Section 2.6, case studies related to real-world attacks and breaches are briefed. Before summarising the chapter in Section 2.8, we discuss the cloud computing vulnerabilities that have to be addressed in Section 2.7.

2.2   DEPLOYMENT MODELS

The deployment models describe the kind of accessibility to a cloud, as well as the location of the cloud. There are four types of models deployed in cloud computing, which are as follows:

- Public cloud
- Private cloud
- Community cloud
- Hybrid cloud.

2.2.1   PUBLIC CLOUD

Public cloud makes it easy for the system and users to be accessible to the general public. Owing to its open system, it could be less secure [9].

2.2.2   PRIVATE CLOUD

Inside the enterprise, private cloud enables the device and the customer to be readily available [10]. Due to its private system, it offers increased security.

2.2.3   COMMUNITY CLOUD

Community cloud allows a certain group of organisations to use its system and services [11].

2.2.4   HYBRID CLOUD

The term "hybrid cloud" is a combination of public and private clouds [12]. A company can use this cloud service application to store confidential consumer data, but link the application as a software service to a public cloud charging application.

2.3   SERVICE MODELS

The cloud infrastructure process is based on service models, which are divided into three categories, which are as follows:

- Infrastructure as a service (IaaS)

- Platform as a service (PaaS)
- Software as a service (SaaS).

2.3.1 INFRASTRUCTURE AS A SERVICE (IaaS)

Many computing services are supported by the IaaS in the context of on-demand servers, operating system, network, hardware, and storage devices. IaaS customers can access the facilities through a wide area network, such as the Internet [13]. For instance, by logging into the IaaS platform, a user can create virtual machines.

2.3.2 PLATFORM AS A SERVICE (PaaS)

PaaS is a platform for developers to write and create their own PaaS [14]. The cloud providers provide a platform including operating system, app development, messaging, database, and a web server. It facilitates rapid growth at an affordable price.

2.3.3 SOFTWARE AS A SERVICES (SaaS)

Instead of downloading and installing an application in a workstation, the user can easily access it through the Internet. This makes the customers free from handling complicated applications and hardware. SaaS consumers are not required to purchase, manage, and upgrade applications or hardware [15]. The only thing needed is to get an Internet connection, and it becomes really easy to access the application.

The most important decisions we need to make when we map our road to the cloud revolve around how much we can and want to handle ourselves versus how much we want the service provider to manage. Here's how technology as a service (IaaS), application as a service (PaaS), and software as a service (SaaS) relate in terms of who handles what. Table 2.1 explores all the possibilities to see which programmes are right and which we want to keep in mind for the future.

2.4 SPECIFIC THREATS IN CLOUD COMPUTING

Due to dynamic scaling, cloud computing has many advantages, including improved speed and stability. Cloud computing, however, has a number of potential drawbacks.

2.4.1 DATA LOSS

Data failure in cloud storage refers to the inability to reach the data stored in the cloud database. Once businesses transfer their processes to the cloud-based environment, the volume of data they have to handle becomes unbearable, rendering backups impractical and costly [16]. Conventional security strategies might not be operable in cloud environments, according to 84% of enterprises. On average, 51% of companies have publicly exposed at least one of their cloud computing services. As a result, there is a risk that unauthorised persons can gain access to data and erase or change an organisation's information. Due to ransomware attacks, which encrypt the

TABLE 2.1

Mapping IaaS, PaaS, and SaaS and Constraints

Services and Technologies	Infrastructure as a Service (IaaS)	Platform as a Service (PaaS)	Software as Service (SaaS)
Applications	Managed by ourselves	Managed by ourselves	Managed by others
Data			Managed by service provider
Run-time		Managed by service provider	
Middleware			
Operating system			
Virtualisation	Managed by service provider		
Servers			
Storage			
Networking			

cloud storage and payment is demanded to retrieve the files, it is required to have an extensive backup, which is one of the most noteworthy threats in cloud computing.

2.4.2 DATA BREACHES AND LEAKS

Among the most frequent threats in cloud environment were data breaches and exposures. Because of the vast volumes of data circulating through staff and cloud services, this occurs [17]. This threat is particularly dangerous in the healthcare profession, where accessibility to up-to-date patient records can potentially make the difference between life and death. Numerous businesses have embraced cloud storage, but they remain unfamiliar with it. It is important to track that their staff use it safely, and it has to be ensured by the organisation.

2.4.3 INSECURE APIs

The key means for interacting with cloud computing networks are application programming interfaces (APIs). API flaws, in particular, will provide hackers with a straightforward route to obtaining customer or workplace privileges [18]. Among the most crucial things to remember is that APIs and user interfaces have always been the most vulnerable aspects of a framework, so it is critical to have security such as API key protection. Users can customise their cloud experience using APIs. However, vulnerable APIs will jeopardise cloud protection and raise the likelihood of being hacked, as illustrated in Figure 2.1.

2.4.4 ACCOUNT HIJACKING

As companies get more reliant on cloud-based computing and software for key enterprise functions, account hijacking is among the most severe cloud protection concerns. Account hijacking is becoming the most serious cloud challenge, as phishing attempts become more successful and targeted [19]. Account hijacking has become

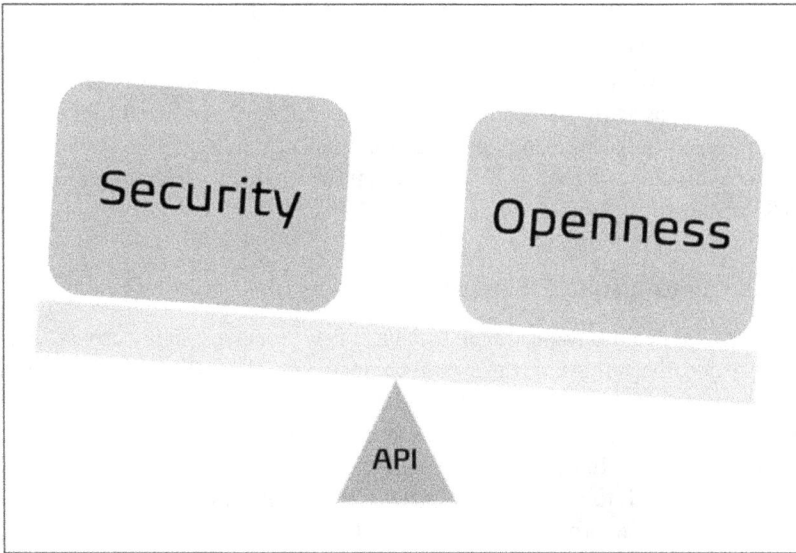

FIGURE 2.1   Security vs. openness.

simple nowadays because many people have extremely weak passwords and reuse their password. Through getting a specific stolen credential being utilised on different accounts, this flaw amplifies the impact of phishing attacks.

2.4.5   MISCONFIGURATION

Cloud data attacks are often caused by misconfigured security settings in the cloud. Most businesses leave their cloud storage's default protection settings unchanged [20]. This may lead everyone to access files easily. A best example for misconfiguration is the National Security Agency's (NSA) poorly configured cloud storage Amazon S3, resulting in the public release of a collection of top-secret records using an unauthorised application. Using a third-party authentication tool would track settings on a regular basis to prevent this. It performs a continuous independent search and notifies you when something is incorrectly installed.

2.5   CLOUD-SPECIFIC VULNERABILITIES

Although most cloud protection experts believe that businesses will profit from the security technologies built into the cloud, they also agree that enterprises will make serious mistakes and reveal sensitive data and systems. Some of the cloud-specific vulnerabilities are described below.

2.5.1   INCOMPLETE DATA DELETION

The user has no information about where the data are actually stored in the cloud space and has less power to check the safe deletion of the information; hence, there are risks involved with data deletion [21]. The above risk is of major concern as the data are shared across a number of different server systems within the cloud service provider (CSP) infrastructure in a multi-tenant environment. Besides that, the retention techniques used by different providers can differ. Firms would be unable to guarantee whether the information is securely deleted; however, no remnants of it should be available to hackers. When an organisation consumes more CSP capital, the risk increases.

2.5.2   COMPROMISED CSP SUPPLY CHAIN

If indeed a CSP decides to outsource sections of its processes, hardware, or services to third parties, those third parties would be unable to follow the CSP's agreed-upon requirements for an organisation [22]. To decide whether the CSP passes its own guidelines to third parties, an organisation must consider how effectively the CSP imposes legislation. If indeed the requirements are not really enforced on the production process, the risk to the company increases. The whole complexity increases as more organisations incorporate many CSP tools because it becomes more dependent on local CSPs and their procurement strategies.

2.5.3   COMPROMISED INTERNET-ACCESSIBLE MANAGEMENT APIS

Clients use the management plane, where the CSPs expose a collection of APIs to manage and interact with cloud services. Firms often utilise APIs to gain accessibility, and to manage, initiate, and monitor the assets and users. APIs for different versions of operating systems, repositories, and many other software may have the same design deficiencies as APIs for many other applications [23]. Unlike management APIs for on-premises storage, cloud service provider APIs are accessible through the Internet. As compliance actors look for vulnerabilities in management APIs, this leaves them more susceptible to exploitation. These bugs can be abused once they are discovered, making a firm's cloud assets vulnerable. Criminals could also exploit the premises of the company to execute strikes on some various CSP customers.

2.5.4   INSIDERS ABUSE AUTHORISED ACCESS

Insiders are certain employees and administrators across both enterprises and CSPs who violate their approved accessibility to a firm's or CSP's networks, systems, and databases, and who are uniquely eligible to cause damage or steal data [24]. Although an intruder inside a company would have the facility to engage in illegal deeds, that forensics can detect the notion of using IaaS is likely to be greater. These forensic techniques cannot be available for cloud providers.

2.5.5   STOLEN CREDENTIALS

Whenever an intruder has exposure to a customer's cloud keys, the intruder will be using the CSP's facilities to gain leverage to remote assets and, if the codes allow for acquisitions, the organisation's assets. The attacker can endanger the company's managerial users, some CSP members, or the CSP's managers using the cloud-based storage space capability [25]. With those rights, a hacker who gains admission to a CSP official's cloud passwords may take advantage of the firm's networks and records. The administrator roles of a CSP and a corporation might be different.

Based on the operation of the CSP's network, the CSP manager has links to the CSP network, programmes, and software, while the customer's managers have access only to the company's cloud computing space. In turn, the CSP manager administers clients of multiple businesses and affords resources for various platforms.

2.5.6 SEPARATION AMONG MULTIPLE TENANTS FAILS

Leveraging interface and program vulnerabilities in a CSP's infrastructure, frameworks, or applications that facilitate multi-tenancy can result in tenants not being kept isolated [21]. Such weakness would be used by an attacker to procure entry into properties or resources attributable towards another entity. If the isolation measures fail, the multi-user environment raises security risks, increasing the possibility of information leaks. One such attack could be carried out by leveraging bugs in the CSP's frameworks, virtualisation layer, or devices attempting to circumvent conceptual abstraction limits, or targeting the CSP's operations API.

2.6 REAL-WORLD ATTACKS AND BREACHES

Cloud computing has changed governments and corporations at an exponential rate, while also posing new vulnerabilities and challenging security threats. The cloud service model has made it possible to provide business-strengthening technologies and enterprises more innovatively than earlier.

The transition from conventional client/server models to service-based models is evolving further by what method information technology departments consider about, build, and provide computational expertise and functions. Fortunately, the increased significance provided by cloud computing developments has resulted in unprecedented security exposures, containing concerns whose full implications are yet uncertain.

This chapter aims to provide companies with the most recent, expert-based knowledge of cloud security risks, challenges, and weaknesses so that the researcher can be informed and make risk management decisions about cloud implementation policies. Thus, the narrations and case reports of leading threats against cloud computing recommend additional data and actionable knowledge. These data can be used to demonstrate how and where these risks work into a broader security picture, as well as how concepts and prevention principles can be applied in real-world situations.

2.6.1 CASE 1: DISNEY PLUS

Disney Plus user accounts are being sold by external cybercriminals in an effort to make money [26]. A synchronous credential stuffing attack resulted in the theft and sale of the majority of such user accounts.

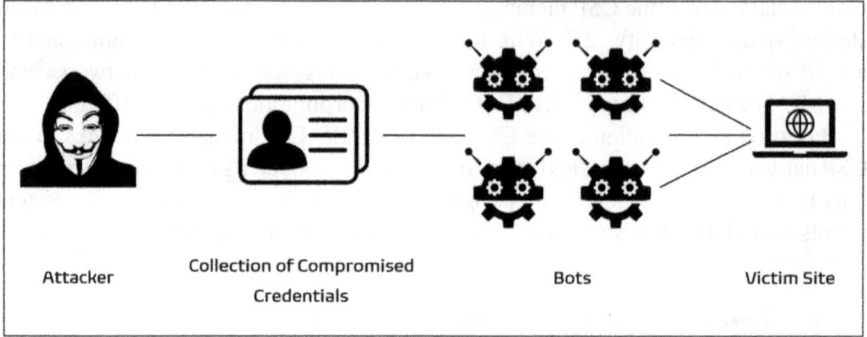

FIGURE 2.2 Credential stuffing.

2.6.1.1 Threat

This sort of attack can be carried out by anyone with a list of stolen credentials, but solitary hackers aren't the real threat. Botnets are a network of linked computers used in large-scale credential stuffing attacks to test thousands of potential username–password combinations as illustrated in Figure 2.2.

2.6.1.2 Impact

As per the documents, hackers gained access to not only the username and password, but also the network and system structure to ensure financial profitability on the acquired accounts, although Disney Plus took some mitigation steps in future.

2.6.1.3 Vulnerability

It has been identified that the vulnerabilities for this attack might be the meagreness of cloud security design and policy, heading towards without getting an event management strategy in place in the event of a breach, a solitary version and passwords for Disney store recreation parks and Disney Plus accounts, inadequacy of uniqueness and credential supervision, lack of inimitable PINs, and lack of multi-factor authentication mandate.

2.6.1.4 Mitigation

A comprehensive security incident management system should be set up before the launch of any service to ensure prompt response and resolution for any possible security incidents, such as account hijacking of newly launched service subscribers. Business continuity plans and organisational readiness should be reviewed at regular intervals to ensure that security incident response plans are effective in the face of real disturbances such as user lockouts, DoS attacks, and other threats. Any flaws or enhancements discovered during the experiments must be incorporated into the company's business continuity and incident management plans.

2.6.2 Case 2: GitHub

Delivery of UDP packets with amplification payload is being used by an anonymous foreign agent to take GitHub offline and disrupt its operations [26].

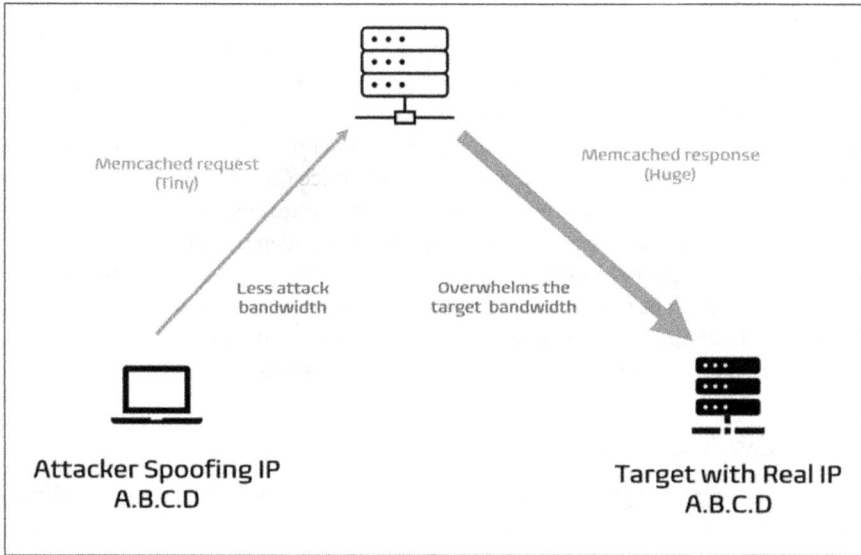

FIGURE 2.3 Memcached DDoS attack.

2.6.2.1 Threat

The actor used a tactic known as Memcrashing to unleash a DDoS attack. Memcrashing operates by exploiting unauthenticated memcached database servers that have been left open to the public Internet. The assault takes advantage of an open device that is connected to the Internet. If an attacker knows the target's IP address, they can spoof it so that any memcached responses go to the target device instead of the attacker's as depicted in Figure 2.3. The DDoS amplification attack works by sending a tiny database request to an open memcached server and setting the source Internet address that GitHub serves in the UDP packet for that message. The memcached database responds with a response that is roughly 50,000 times the size of the command – a 203-byte request results in a 100 MB response. Inbound traffic flow on GitHub increased at 1.35 terabits per second, or 126.9 million packets per second. The overwhelming amount of data overloaded GitHub's machines, forcing them to shut down and stop listening to regular users.

2.6.2.2 Impact

The system and its data were inaccessible for ~5 minutes, which was insufficient for GitHub to report any material impact. If the system had been down for a longer time, GitHub, a shared software development site, may have been forced to postpone a variety of projects.

2.6.2.3 Vulnerability

Inadequate access management, misconfiguration, open ports, failure of cloud protection architecture, poor architectural structure, out-of-date software, and insufficient training for insiders have all been described as possible vulnerabilities for this attack.

2.6.2.4   Mitigation

There should be a mechanism for calculating the effect of any damage to the enterprise that is described and registered (cloud provider and cloud consumer). High-level risk conditions and information streams that may have regulatory policy implications must be specifically identified in network engineering diagrams. MAC spoofing and ARP poisoning attacks, as well as distributed denial-of-service (DDoS) attacks, are examples of network-based attacks that incorporate anomalous intrusion or exit traffic patterns and can be detected and responded to using defence-in-depth strategies such as deep packet inspection, traffic throttling, and blackholing. Traffic between trustworthy and untrusted links must be restricted and monitored in network environments and simulated instances. All set-ups must be checked at least once a year and backed up with documentation for all allowed services, protocols, and ports, as well as compensating controls.

2.6.3   CASE 3: ZOOM

The event of uninvited groups hijacking video conversations to interrupt the ordinary proceedings is known as Zoombombing [26]. Zoombombers mine 500 million user accounts and hack private information due to Zoom's lack of security protections.

2.6.3.1   Threat

Attackers aim to "stuff" existing usernames and passwords into the login pages of other digital resources using a database of known usernames and passwords. Attackers may also use one piece of login information to open several accounts because users reuse their credentials across multiple domains.

2.6.3.2   Impact

Attackers splatter racially incendiary and sexually suggestive vandalism on business and school meetings.

2.6.3.3   Vulnerability

With the COVID-19 pandemic, Zoom saw a massive increase in users, with many incidents in early 2020. Several problems emerged, including meeting room details that were incorrectly randomised, quickly guessed, or broadly broadcast without adequate detective or protective security measures. Without proper Zoom protection measures, customer credential reuse was rampant. Finally, attackers may exchange links that leak Windows network passwords using the Zoom Windows client's group chat functionality. As Zoom transforms Windows UNC paths into clickable connections, this occurs.

2.6.3.4   Mitigation

Since Zoom is a tech utility, hackers are more likely to attack older versions. Threat modelling can be used to design technical mechanisms such as security and privacy controls, not showing meeting information, openly and using proper random numbering sequences. Threat research revealed no vectors involving uninformed workers, allowing for strict protocol enforcement and device settings for meeting passwords, no enter before host, member screen sharing, and locking the meeting after it begins.

2.6.4 GENERIC THREATS AND VULNERABILITIES

Some generic threats and vulnerabilities are listed below.

2.6.4.1 Threats

- Amplification attack is defined as any attack in which an attacker consumes more resources than a single link can [27]. The amplification effect increases the attack's intensity by using asymmetry, under which a lack of resources results in a high number of goal failures.
- Memcrashing is a form of amplification attack [28] that uses a flaw in the memcached registry on UDP port 11211 to paralyse the hosting server. Client-side libraries are used by memcached clients to communicate with servers on port 11211. Memcached servers display their support on both TCP and UDP port 11211 by contrast.
- Cloud server abuse and nefarious use: VPN and anonymous networking platforms are used to exploit identities.
- Insiders – employees, contractors, and those with access privileges that have been poorly conditioned to challenge or ignore potentially malicious email.

2.6.4.2 Vulnerabilities

- An intruder tricked a server into granting access to cloud server settings, including keys to everything the server has access to, due to a server-side request forgery (SSRF) vulnerability [29] on the network.
- One of their certified and probably trustworthy security providers could not have password-protected the cloud storage.
- Internal design and human error on the part of an internal cloud team introduced server weakness (undisclosed server vulnerability) and conditions that enabled the cyberattack, deliberately making the server Internet-reachable (misconfiguration and inadequate change control) and establishing access to the business database images from this server via cloud API access keys (absence of cloud security architecture and strategy) [26].
- Intimate information is needed for proper deployment and configuration decisions in a complex environment.
- Users were not given any substantive updates regarding the feature, nor did they have the option to agree to data collection and conversion.
- The web application's data are stored on a public server, and there is no security scheme to access it. During the data migration process, there were insufficient security checks. Inadequate supervision and review mechanisms are examples of poor vendor risk management practices.
- Inadequate credential control and successful encryption mechanisms could have allowed lateral network travel.
- Insufficient antimalware and security testing failed to identify and prevent mining scripts from being installed.

2.7 CLOUD COMPUTING VULNERABILITIES TO BE ADDRESSED

When planning to move to the cloud, we must keep the following cloud vulnerabilities in mind.

2.7.1 PAUCITY OF CONTINUOUS SCANNING

Since applications are continually being turned up and down, clients are often unaware of new products in their system, and rapid implementation will lead to rapid introduction of issues. Companies have found it difficult to track and orchestrate protection around applications due to the simplicity of which they can be integrated into an environment [30]. Clients frequently have several teams launching cloud services, making it impossible for businesses to handle them centrally when they are not really aware of the situation. Corporations ought to scan their servers on a regular basis to guarantee that all data are encrypted and there are no backdoor server versions available. Too frequently, the encryption algorithms in use are insecure, putting SSLs, servers, and serverless environments at risk.

2.7.2 DEARTH IN ADHERENCE TO POLICIES

Frequently, organisations carve cloud protection policies in a manual and assign it to the DevOps unit [30] devoid of really understanding how well these strategies will be implemented. Everything in the industry is expected to configure and execute cloud protection policies, according to security departments; DevOps teams, however, anticipate a virtual script or some other automatic solution to do it for them. To access APIs, account identification is required; since most businesses provide a slew of separate credit reporting agencies, it's difficult to know whom to contact. Organisations struggle to comprehend the cloud and wrap their arms around it because of the difference between written legislation and resources to simplify the development process. Given how often server misconfigurations occur, the danger is that departments are spinning up insecure cloud systems when no one else in the company is conscious of the risks involved.

2.7.3 BREAKDOWN IN THE MODEL OF SHARED RESPONSIBILITY

Adversaries are taking advantage of a breakdown in the mutual responsibility paradigm when it comes to data ownership privileges and data standards [30]. Clients are granted access to data based on their job title or responsibilities in a standard structured system, and the administrator may track, retain, manipulate, or administer their access. Data discovery, classification of data, as well as ongoing review, and validation of an entity sought to access the information should all be continued when transitioning to a cloud world. During the relocation process, it's easy to forget the oversight and compliance of individual access rights. Market and financial decision-makers usually facilitate and conduct migrations outside of the security unit, and security professionals are increasingly active in the process. As a result, after the transition has been completed, and daily operations have resumed, data access problems can arise.

2.7.4 Absence of Security around Databases

As they race to market with resources that could be more experiment than manufacturing, many companies leave default database configurations in place [30]. When businesses compete for cloud market share, they don't need third-party approval for code
updates and don't check deployments for errors until they've been created. The database defaults or the person who created the database to protect the instance might not
be adequately hardened. To ensure that all database incidents have been found and shut
down, organisations should have familiarity with database applications. Firms need to
ensure that only customers who require unencrypted access to data are granted access
in order to comply with PCI or CCPA. The most important thing for businesses to do
is to ensure that mechanisms to check for anticipated activity are in effect. Any asset
losses or data theft would be flagged, which would help in emergency preparedness.

2.7.5 Low-Entry Barriers for Bad Actors

The cloud's pervasive and available computing and storage capabilities have ensued
in low-entry impediments for those seeking to make use of cloud environments for
nefarious reasons [30]. To begin with, the cloud would be used as a launch pad for
incidents because it gives attackers a relatively anonymous environment in which to
organise nefarious activity that is simple to set up or break down. Next, the hackers
will be able to quickly set up or blackout the cloud compute capacity in the cloud,
making it open to malicious operations ranging from DDoS attacks to phishing
attacks using host networks or device services. Threat actors see workers with substantial server responsibilities or main entrance as a desirable attack option. Finally,
in public or hybrid cloud environments that are less stable than expected, hackers
may hijack access control transactions or flows.

2.7.6 Automated Attacks

As tech becomes less costly, attacker increases efficiency. As users connect more systems to the Internet and technologies become more diverse, automated cloud attacks
have become easier and more frequent [30]. The potential for any developer to bring
a device online with only a credit card swipe has opened the door to heavily lever-
aged automated attacks. Large-scale data analysis and collection has become simpler for rivals, thanks to software advances, and there are frequently inconsistencies
in how a policy manual states it can work and how things actually work. In order
to defend from automated threats, enterprises should determine where all of their
devices are, what really is running on it, and what standard they use. Businesses must
now embrace robotics for their security, using software engineering skills to simplify
the segregation and restraint future accidents, giving people more time to investigate.

2.8 CONCLUSIONS

In preparation for an enterprise organisation to effectively migrate their existing
operations to the cloud, they must be mindful of the cloud risks. We ought not depend

on the cloud service provider to keep hold of protection for us; instead, we can learn about security risks and vulnerabilities and consult with our CSP and find out how they're dealing with them and proceed from there. Regardless of whether the CSP currently has backup services, we can build remote backups of our data and it is easier to have multiple different backups than to discover that the data were not backed up at all when the necessary data restoration occurs.

Using real-world attacks and hacks, the case study series presented in this chapter aims to draw the dots between risks when it comes to security research. The illustration is described as a structured narrative that includes an attack-style overview of the perpetrator, including risks, vulnerabilities, and endpoint measures and mitigation strategies. Researchers and engineers are encouraged to use these data as a reference point for their own research and measurements. Extended perspectives of the case scenarios facilitate potential details, including how an event occurred or how it could have been handled, as well as guidelines for more study. This chapter inferred predicted findings in the scenarios where facts such as influences or mitigations were not shared previously. The researcher will find this initiative beneficial for their research inclusions in future publications.

REFERENCES

1. M. G. Avram, "Advantages and challenges of adopting cloud computing from an enterprise perspective," *Procedia Technol.*, vol. 12, pp. 529–534, 2014, doi: 10.1016/j. protcy.2013.12.525.
2. A. Ghazizadeh, "Cloud computing benefits and architecture in E-learning," *In Proceedings of the 2012 17th IEEE International Conference on Wireless, Mobile and Ubiquitous Technology in Education, WMUTE 2012*, 2012, pp. 199–201, doi: 10.1109/WMUTE.2012.46.
3. F. F. Moghaddam, M. B. Rohani, M. Ahmadi, T. Khodadadi, and K. Madadipouya, "Cloud computing: Vision, architecture and characteristics," *In Proceedings of the 2015 6th IEEE Control and System Graduate Research Colloquium, ICSGRC 2015*, February 2016, pp. 1–6, doi: 10.1109/ICSGRC.2015.7412454.
4. H. Suo, Z. Liu, J. Wan, and K. Zhou, "Security and privacy in mobile cloud computing," *In Proceedings of the 2013 9th International Wireless Communications and Mobile Computing Conference, IWCMC 2013*, 2013, pp. 655–659, doi: 10.1109/IWCMC. 2013.6583635.
5. T. M. C. Simoes, J. J. P. C. Rodrigues, J. E. F. Costa, and M. L. Proenca, "E-learning solutions for cloud environments," *In Proceedings of the 2012 IEEE Latin America Conference on Cloud Computing and Communications, LatinCloud 2012*, 2012, pp. 55–59, doi: 10.1109/LatinCloud.2012.6508158.
6. M. Adimoolam, A. John, N. M. Balamurugan, and T. Ananth Kumar. "Green ICT communication, networking and data processing." In B. Balusamy, N. Chilamkurti, and S. Kadry (eds.) *Green Computing in Smart Cities: Simulation and Techniques*, pp. 95–124. Springer, Cham, 2021.
7. N. Roman Herbst, S. Kounev, and R. Reussner, "Elasticity in cloud computing: What it is, and what it is not," *In the Proceedings of the 10th International Conference on Autonomic Computing, ICAC 2013, June 26-28, San Jose, CA*, 2013.
8. P. Mell, "What's special about cloud security?" *IT Prof.*, vol. 14, no. 4, pp. 6–8, 2012, doi: 10.1109/MITP.2012.84.

9. K. Xue and P. Hong, "A dynamic secure group sharing framework in public cloud computing," *IEEE Trans. Cloud Comput.*, vol. 2, no. 4, pp. 459–470, 2014, doi: 10.1109/TCC.2014.2366152.

10. R. Basmadjian, H. De Meer, R. Lent, and G. Giuliani, "Cloud computing and its interest in saving energy: The use case of a private cloud," *J. Cloud Comput.*, vol. 1, no. 1, pp. 1–25, 2012, doi: 10.1186/2192-113X-1-5.

11. S. Bruque-Cámara, J. Moyano-Fuentes, and J. M. Maqueira-Marín, "Supply chain integration through community cloud: Effects on operational performance," *J. Purch. Supply Manag.*, vol. 22, no. 2, pp. 141–153, 2016, doi: 10.1016/j.pursup.2016.04.003.

12. Q. Li, Z.-yuan Wang, W.-hua Li, J. Li, C. Wang, and R.-yang Du, "Applications integration in a hybrid cloud computing environment: Modelling and platform," *Enterp. Inf. Syst.*, vol. 7, no. 3, pp. 237–271, 2013, doi: 10.1080/17517575.2012.677479.

13. A. P. Rajan and S. Shanmugapriyaa, "Evolution of cloud storage as cloud computing infrastructure service," *IOSR J. Comput. Eng.*, vol. 1, no. 1, pp. 38–45, 2013. Accessed: April 21, 2021. [Online] Available: http://arxiv.org/abs/1308.1303.

14. A. Giessmann, K. Stanoevska-Slabeva, B. Rajagopalan, C. Magnusson, and G. Juell-Skielse, "Business models of Platform as a Service (PaaS) providers: Current state and future directions," *J. Inf. Technol. Theory Appl.*, vol. 13, no. 4, pp. 31–55, 2012.

15. V. V. Glukhov, I. V. Ilin, and O. J. Iliashenko. "Improving the efficiency of architectural solutions based on cloud services integration." O. Galinina, S. Balandin, and Y. Koucheryavy (eds.) *In Internet of Things, Smart Spaces, and Next Generation Networks and Systems*, pp. 512–524, Springer, Cham, 2016.

16. M. Kazim and S. Ying, "A survey on top security threats in cloud computing," *Int. J. Adv. Comput. Sci. Appl.*, vol. 6, no. 3, 2015, doi: 10.14569/IJACSA.2015.060316.

17. L. Cheng, F. Liu, and D. D. Yao, "Enterprise data breach: Causes, challenges, prevention, and future directions," *Wiley Interdiscip. Rev. Data Min. Knowl. Discov.*, vol. 7, no. 5, p. e1211, 2017, doi: 10.1002/widm.1211.

18. N. Khan and A. Al-Yasiri, "Identifying cloud security threats to strengthen cloud computing adoption framework," *Procedia Comput. Sci.*, vol. 94, pp. 485–490, 2016, doi: 10.1016/j.procs.2016.08.075.

19. A. Mirian, J. De Blasio, S. Savage, G. M. Voelker, and K. Thomas, "Hack for hire: Exploring the emerging market for account hijacking," *In The Web Conference 2019 Proceedings of the World Wide Web Conference, WWW 2019*, May 2019, pp. 1279–1289, doi: 10.1145/3308558.3313489.

20. R. Sulatycki and E. B. Fernandez, "Two threat patterns that exploit 'security misconfiguration' and 'sensitive data exposure' vulnerabilities," *In ACM International Conference Proceeding Series*, July 2015, vol. 08, pp. 1–11, doi: 10.1145/2855321.2855368.

21. M. T. Khorshed, A. B. M. S. Ali, and S. A. Wasimi, "A survey on gaps, threat remediation challenges and some thoughts for proactive attack detection in cloud computing," *Future Gener. Comput. Syst.*, vol. 28, no. 6, pp. 833–851, 2012, doi: 10.1016/j.future.2012.01.006.

22. O. Akinrolabu, S. New, A. Martin. "Cyber supply chain risks in cloud computing: Bridging the risk assessment gap - ORA - Oxford University Research Archive." 2017. https://ora.ox.ac.uk/objects/uuid:751fc4ce-1cfb-45f9-b442-d6c76f099076 (accessed April 21, 2021).

23. H. Shah, S. S. Anandane, and S. Shrikanth, "Security issues on cloud computing," *Handb. Res. Secur. Considerations Cloud Comput.*, pp. 1–29, 2013. [Online] Available: http://arxiv.org/abs/1308.5996 (accessed April 21, 2021).

24. D. N. Le, C. Bhatt, and M. Madhukar, *Security Designs for the Cloud, IoT, and Social Networking*. Hoboken, NJ: Wiley, 2019.

25. K. Thomas et al., "Data breaches, phishing, or malware? Understanding the risks of stolen credentials," *In Proceedings of the ACM Conference on Computer and Communications Security*, October 2017, pp. 1421–1434, doi: 10.1145/3133956.3134067.

26. CSA. "Top threats to cloud computing: Egregious eleven deep dive I CSA." 2020. https://cloudsecurityalliance.org/artifacts/top-threats-egregious-11-deep-dive/ (accessed April 21, 2021).

27. S. Dong, K. Abbas, and R. Jain, "A survey on Distributed Denial of Service (DDoS) attacks in SDN and cloud computing environments," *IEEE Access*, vol. 7, pp. 80813–80828, 2019, doi: 10.1109/ACCESS.2019.2922196.

28. Security Boulevard. "Massive DDoS attack washes over GitHub: Security Boulevard." 2018. https://securityboulevard.com/2018/03/massive-ddos-attack-washes-over-github/ (accessed April 21, 2021).

29. B. Jabiyev, O. Mirzaei, A. Kharraz, and E. Kirda, "Pre-venting server-side request forgery attacks," 2021, doi: 10.1145/3412841.3442036.

30. CRN. "12 biggest cloud threats and vulnerabilities in 2020." 2021. https://www.crn.com/slide-shows/security/12-biggest-cloud-threats-and-vulnerabilities-in-2020 (accessed April 22, 2021).

3 Security and Privacy Provocation of Data in Cloud Computing

D. Prabakaran, K. Suresh Kumar,
and R. Senthil Kumaran
IFET College of Engineering

CONTENTS

DOI: 10.1201/9781003219880-3

3.1   CLOUD COMPUTING: AN INFRASTRUCTURELESS PARADIGM

The term "big data" refers to a set of data of a larger size that can be output by a specific program. It can also refer to a large variety of data types, and the datasets are too large to peruse or query on a regular computer. The term "cloud" refers to the mechanism that takes the data in and does any operations on the corresponding data. Cloud computing supports employing a "software as a service" (SaaS) model to process the data to the customers. Here, all the processes can be done from the user site interface. Big data can be in any format; that is, it can be in any standard or non-standard format. If it is in a non-standard format, the data can be converted into a standard form through artificial intelligence and machine learning support.

Nowadays, cloud computing is involved in all the sectors, which include banking, retail, real estate, health, telecom, and IT industry. Most of the activities of e-commerce use many cloud applications. Based on the research of Prof. Luis M. Vaquero et al., in the year 2008 alone, 22 various definitions were found. One of the aspects of computing is that the whole IT industry is renovating from a corporal world towards a simulated world. Joshi et al. explained that the cloud consists of a group of computers interconnected in a virtual manner, which is established through intervention across the customers and the service providers [1]. The cloud computing aims to offer the services with determining the storage, network, software, and

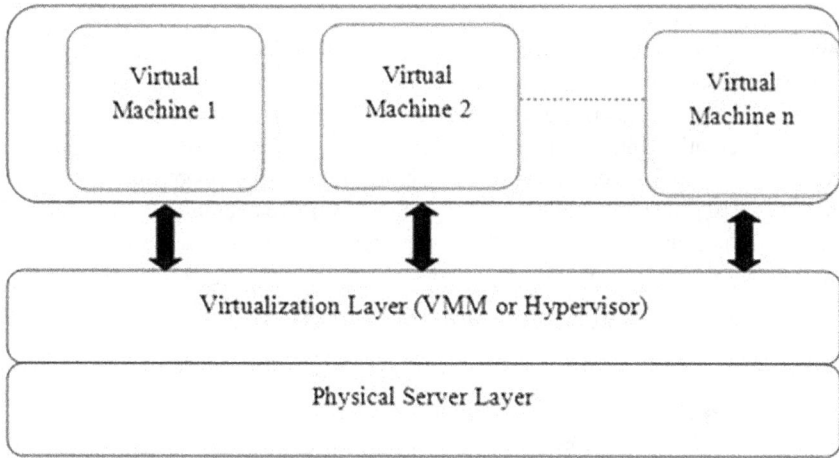

FIGURE 3.1 Layered virtualization technology architecture.

a service cross those combinations. The three significant taxonomies involved in the service of cloud computing are infrastructure, platform, and software as a service, ranging from simulated servers to several applications. Cloud computing offers infrastructure as a service (IaaS) similar to pay as you use for the end-users. For platform as a service (PaaS), Aneka is a type of platform that integrates the private and public clouds. Cloud architecture is an autonomic cloud engine that uses a self-organizing overlay and supports flexible content-based routing (Figure 3.1).

3.2   PROCESS OF MIGRATION INTO A CLOUD

The process of migration involves seven stages.
- Conducting cloud migration assessments
- Isolating the dependencies
- Mapping the messaging environment
- Re-architecting and implementing the lost functionalities
- Leveraging cloud functionalities and features
- Testing the migration
- Iteration and optimization.

The first step of this seven-step model is assessment. This step is to realize the specific issues within the particular case of migration at the application range, strategy, and architecture or procedure levels. This can be done for the tools used and the functionalities [2] in the enterprise presentation. This stage includes recurring cost, migration cost, database segmentation, migration, and NFR support. The second stage is meant for isolating the dependencies. The following processes are made on this level: run-time environment, licensing, library, architectural, and application dependencies. During the mapping process, messages are mapped into marshalling and demarshalling messages. Also, run-time and library approximations are made. At the re-architect level, this approximates the lost functionality using cloud run-time support. The fifth level of migration supports exploiting additional cloud features and seeks low-cost augmentations, auto-scaling, storage, bandwidth, and security. During the testing phase of the migration strategy, new test cases due to cloud augmentation and tests for production loads have been carried out with proof of concept. In the last phase, the various processes involved are to optimize rework and iteration, satisfy the cloudonomics of migration, optimize compliance with standards and governance, and develop a roadmap for leveraging new cloud features[3].

3.3   CLOUD COMPUTING CHARACTERISTICS

The general characteristics of cloud computing is as follows:
- Resource pooling
- On-demand self-service
- Easy maintenance
- Large network access
- Availability
- Automatic system
- Economical
- Security
- Pay as you go
- Measured service.

3.3.1   RESOURCE POOLING

One of the major characteristics of cloud computing is resource pooling. Different services can be provided to numerous clients through the resources shared among the cloud service providers. Various types of services such as bandwidth allocation, data storage, and processing are supported by different clients. In real time, allocation of resources in the administration procedure doesn't have to come into contact with the client.

3.3.2   ON-DEMAND SELF-SERVICE

This is the fundamental characteristic of cloud computing. The server on time, skills, and storage capability can be monitored by the clients.

3.3.3   EASY MAINTENANCE

The servers can be maintained effectively so that they will not be down at any time. The updates can be frequently done to note the capabilities of the client.

3.3.4   SCALABILITY AND RAPID ELASTICITY

Rapid scalability means more servers can be operated simultaneously, i.e. more workloads from the client side simultaneously. Clients have workloads that can run cost-effectively owing to the rapid scalability.

3.3.5 ECONOMICAL

The cloud system reduces an organization's IT expenditure. The client has to bear the amount for the space used for administration. It is to ensure that no extra charge has to be paid. Some free space can also be allotted for clients.

3.3.6 MEASURED AND REPORTING SERVICE

The major cloud characteristic that has the finest choice for organizations is reporting services. This service is very much useful for cloud service providers and clients. The services and the purpose for the client and the service provider can be monitored and reported. The resource usages and the billing can also be monitored (Figure 3.2).

3.3.7 SECURITY

It is essential to secure the data used in cloud servers. To prevent data loss, cloud services can take a duplicate copy of the data and store it. In case if a server loses the

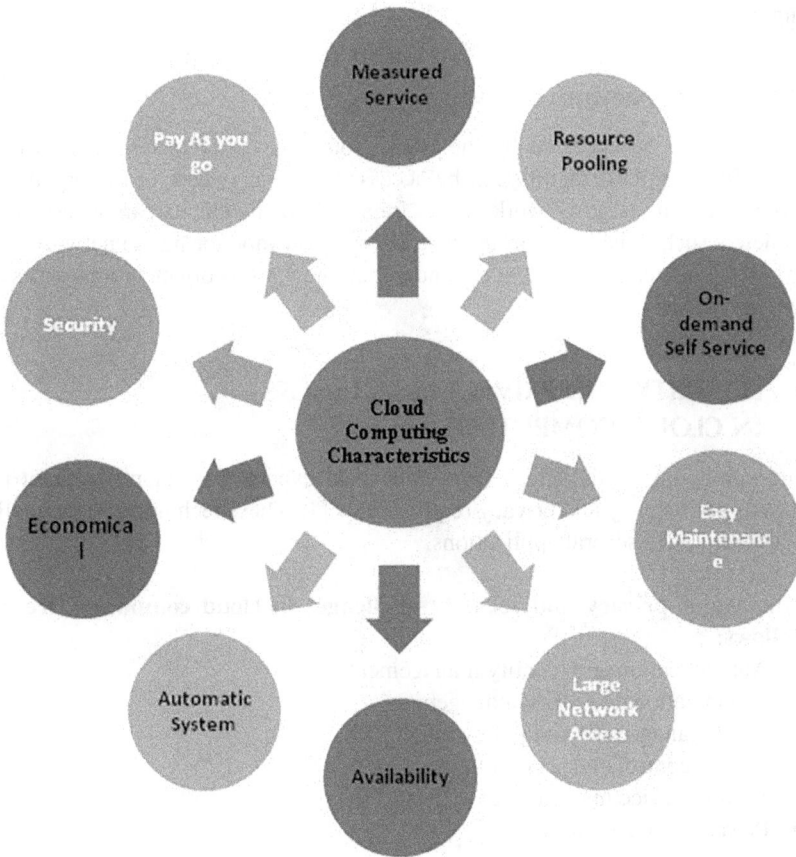

FIGURE 3.2 Characteristics of cloud computing.

data, it can be restored from another server. This can be applicable during real-time, simultaneous access of a specific file by more number of users, which may lead to corruption of the file.

3.3.8 AUTOMATION

Cloud computing can have the ability to automatically install, configure, and maintain a cloud service. This reduces the manual process. Even though it reduces manual efforts, it is difficult to achieve automation in the cloud system. It needs virtual machines, servers, and large storage for deployment.

3.3.9 RESILIENCE

Resilience is defined as the service that is able to recover from any odd state to its original state. In cloud resilience, this can be referred to as how the servers, databases, and systems recover to the original state from the affected state. Due to the remote access of cloud servers, there are no restrictions and limitations in using it [4].

3.3.10 LARGE NETWORK ACCESS

Cloud data can be accessed by the client from faraway countries using a device and the Internet. This facility can be accessed from anywhere in the organization via Internet. This large network access implies how the clients can access various parameters such as latency, throughput, and delay by monitoring the network.

These characteristics make it a standard among various organizations across different industries.

3.4 SECURITY AND PRIVACY CHALLENGES IN CLOUD COMPUTING

Cloud computing methodology uses multi-field expertise to adopt different techniques as per security and privacy requirements [4]. These techniques signify all the services related to several applications.

The major privacy and security challenges in cloud computing are as follows:
- Authentication and identity management
- Organizational security management
- Trust management and policy integration
- Access control and accounting
- Secure service management
- Privacy and data protection
- Access control needs
- Secure interoperation.

3.4.1 AUTHENTICATION AND IDENTITY MANAGEMENT (IDM)

This helps to validate the users, the credentials, and characteristics-based services. The significant issue here is in identifying the token and in protocol negotiation. This identity management system [5] must protect the user- and process-related sensitive information. Some other security-related components can also be integrated with this authentication and IDM.

3.4.2 ORGANIZATIONAL SECURITY MANAGEMENT

Based on external entities, the reaction to security events and also for a recovery plan has always caused fear in this scenario. The clients always consider the new risks such as the data outflow within multi-tenant clouds and local disaster problems. Best practices and standards have to be re-evaluated to adapt in clouds for security.

3.4.3 TRUST MANAGEMENT AND POLICY INTEGRATION

In cloud, many service providers are providing different security services and mechanisms. Hence, it is essential to consolidate the similarity from their policies. Due to various services, a framework can be adopted for consolidating some parameters in order to establish trust for sharing the needs.

3.4.4 ACCESS CONTROL AND ACCOUNTING

Access control services have to combine the requirements of privacy shield through specific protocols. This can be offered by the cloud models for appropriate access control and framework for accessing issues in the cross-domain [6].

3.4.5 SECURE SERVICE MANAGEMENT

In order to optimize the resource allocations, the service provider uses virtualization technologies that separate application services and infrastructure. This needs automatic service and composition frameworks to describe specific standards for secure operation.

3.4.6 PRIVACY AND DATA PROTECTION

In the cloud, data can be shared with all, but only the data owner can edit the data. A standard data-centric approach is needed to move the data protection from applications and servers. If any client needs data access, the server can verify the policies and reveal only when the procedures are acceptable. Older techniques are used for securing the data, but attention is needed for privacy and computation.

3.4.7 ACCESS CONTROL NEEDS

For access control, several methods have been proposed so far. Due to the simplicity and flexibility in a dynamic environment, role-based access control (RBAC) is accepted as the favourable model. One of the other advantages of RBAC is that it can be suited for the integration of various policy needs. In cloud servers, the users are not known by the service providers initially; hence, assigning users is a very much challenging task in access control policies.

3.4.8 SECURE INTEROPERATION

To access the various policies from various domains and in global-level policies, secure interoperation and policy engineering mechanism are addressed by multiple researchers. The dynamic environment, transient domains and, for a specific purpose, interaction make it inappropriate for the centralized and decentralized schemes [7].

3.5 CASE STUDY ON VARIOUS ATTACKS IN CLOUD COMPUTING

Cloud computing attracts users, and in the same way, it attracts more hackers to invade its sensitive resources illegally. Cloud computing is prone to numerous kinds of attacks as aforementioned. Although establishing a variety of security features, the hackers tend to succeed at certain times in their attempt to access cloud data resources illegally. There are many examples of these unauthorized attacks in cloud computing, and this session throws light on certain notable attacks in cloud computing.

3.5.1 FINTECH COMPANY – JULY 2020

The FinTech Company accepted that their data related to 7,516,625 customers were breached through their login credential confirmation in an online sale launched at a public forum. The company pinpointed their technical team for this data breach and had exposed data such as user name, contact number, address, mailing id, date of birth, and so on. Not restricted to these data, sensitive financial data such as credit card numbers and band account details were also breached during this cyberattack.

3.5.2 AMBRY GENETICS DATA BREACH – APRIL 2020

The Ambry Genetics Company issued a public notice on April 2020 stating that their IT professional team has diagnosed a data attack through illegal access to an employee's mail account. The analysis report states that poor authentication policy and credentials were the platform to host this attack. The hacker launched an impersonation attack to identify the credential pattern of the authorized user and succeeded in accessing the victim's mail account pretending to be an authorized user. Based on the analysis report, security policies were redefined to create a strong basis for mail security by the organization.

3.5.3 EQUIFAX DATA BREACH – JULY 2017

Equifax is a credit reporting agency with an aggregate of 800 million data related to customer information and owns over 88 million businesses worldwide. During July 2017, the IT security [8] professional of the organization deduced a blocked and suspicious activity in the organization's network. On analysing the network pattern, these suspicious activities were found to be related to the customers in USA. The organization was alarmed by the Apache Struts vulnerability earlier in March 2017. In turn, Equifax failed to strengthen the security policies of the organization, which led to this successful attack in the month of July. The hackers accessed the database illegally and acquired access to dozens of database accounts and their details in addition to the creation of backdoors in the organization's network.

3.5.4   Uber Data Breach – November 2017

Uber, the well-known online portal for transport hiring, announced that their database was breached on November 2017. The announcement further provided details that the hacker attacked the Uber server which stores the customer's personal information including the location information of the user and demanded an amount of $100,000 to vanish the copy of the data from the hacker's end. The hackers accessed a private GitHub repository that was utilized by Uber to store the riders' data. The analysis report states that the Uber database had a poor cover-up, and the security policies were very poor in maintenance.

3.6   ANALYSIS OF CASE STUDIES

The case studies, as mentioned earlier, are the bread pieces of the hacker's activities, where thousands of attacks are performed successfully with the involvement of advanced technologies. These attacks aim at gaining ransom directly or indirectly from the authenticated party. The analysis of case studies related to attack incidents [9] yields certain weaker network policies. Some of the reasons and thrust areas to concentrate much for the improvement of security policies [10] are as follows:

* Cyber espionage
* Web vandalism
* Distributed denial of service (DDoS)
* Infrastructure attacks
* Hardware compromise – theft and damage.

3.6.1   Cyber Espionage

Cyber espionage is a common method launched by hackers to access data illegally. The information retrieved by the hackers illegally may or may not possess the required information [11]. Rather, the information retrieved may be used for blackmail purposes demanding ransom from the victim.

3.6.2   Web Vandalism

An attacker tends to launch web vandalism by disfiguring the targeted webpage and modifying the contents and policy principles of the meagre secured framework.

In digital vandalism, the authenticity of the webpage is compromised to edit the contents without proper authentication. One of the famous examples for web vandalism is that the NATA attacked the terrorist background website to edit the content.

3.6.3 DISTRIBUTED DENIAL OF SERVICE (DDoS)

The denial of service [12] is a well-known attack used by most hackers. In this type of attack, the hackers tend to send multiple numbers of request packets to the server from a single or multiple devices such that the server will be overwhelmed with the repeated requests. Due to the overwhelming of repeated requests, the server cannot be able to accept genuine requests from authenticated users and will remain unavailable to them. This kind of attack is termed as the non-availability of resources to authorized users. This attack has more possibility of damaging the hardware infrastructure, leading to permanent non-availability of the services.

3.6.4 INFRASTRUCTURE ATTACKS

The infrastructure attack is a process of harming public properties by injecting harmful network contents such as a virus – malware intended to vanish or interrupt the routing functionalities. This attack can be overcome with the utilization of proper intrusion detection and prevention systems such as anti-virus scanning for virus and malware.

3.6.5 HARDWARE COMPROMISE: THEFT AND DAMAGE

The attackers prefer to establish a physical connection with the hardware rather than hosting software attacks. These attacks involve a much complicated process to break the framework depending on the rigidness of the algorithm. The hacker, when succeeded in accessing the hardware physically, may tend to damage the hardware or may steal the hardware to access the database.

3.7   CLOUD COMPUTING ATTACKS: A STATISTICAL ANALYSIS

A statistical analysis of attacks in cloud computing was performed to yield the quantitative analysis and identify the weaker sections of cloud computing security policies. On performing global analysis on cloud computing attacks, it is seen that the count of attacks drastically increased with technology and algorithms. Figure 3.3 depicts the attack counts in millions that have been recorded every year. The analysis report states that the count increases drastically, not exponentially demanding ransom from the victim, and the reason behind the increase in count refers to the poor security policies followed in cloud data centres.

The depth analysis on this statistics of the attacks in cloud computing and its resources yields additive information regarding the types of attacks, and the techniques are depicted in Figure 3.4.

Figure 3.4 shows the percentage level of cloud computing attack techniques as per the analysis done in November 2020. The major contribution of 27.4% was performed

No. of Cyber Attacks (In Millions)

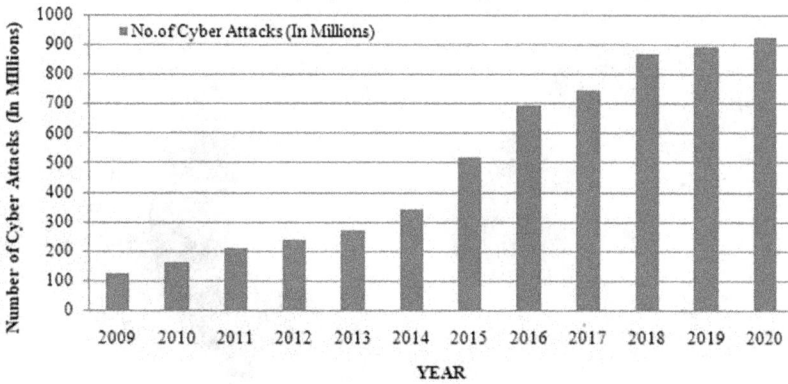

FIGURE 3.3 Cloud computing attacks recorded year-wise.

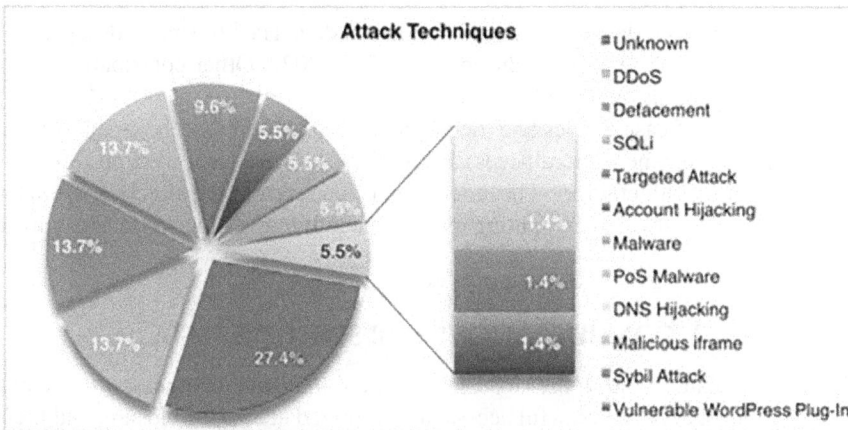

FIGURE 3.4 Cloud computing attack techniques.

by various new technology attacks such as ransomware attacks [13] and PUA attacks. The next most frequent attacks are the denial of service, defacement attack, and SQL injection attack, contributing to 13.7% each. Apart from these attacks, specific attacks [14] such as account hijacking, malware, Sybil attack, WordPress attacks, POS malware, and DNS hijacking contribute well effectively, and other attacks make the security of cloud computing highly vulnerable. The motivation behind this increase in cloud computing attacks is analysed, and it is found that the ransom demanding is not the only ultimate task of the hackers. Figure 3.5 mentions the motivation of the attackers in launching cloud computing attacks.

The analysis results depicted in Figure 3.5 mention that cybercrime is the ultimate aim of hosting cloud computing attacks with a maximum of 67% and it has increased

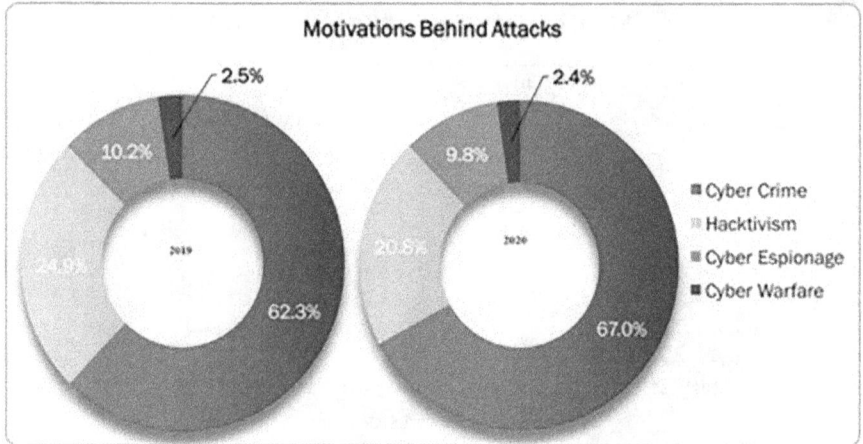

FIGURE 3.5   Motivation behind cloud computing attacks.

when compared with the previous year's statistical data. Hacktivism holds a percentage of 24.9% in 2019 and has reduced to 20.8% in 2020. Other contributions were held by cyber espionage and cyberwarfare.

The analysis of case studies and the attack analysis reveal that the cloud computing attack has become a more necessary concern to be concentrated on to reduce cloud computing vulnerability. The reduction in cloud computing attacks leads to the improvised usage of cloud computing resources and improves the world economy to a greater extent.

3.8   PRIVACY CONSIDERATIONS FOR SENSITIVE DATA

The sensitive data stored in the cloud storage are vulnerable to attacks with the aim of unintentional access, unlawful access, unauthorized access, disclosure, and theft. The confidentiality of data deals with preserving the privacy of the information that frames the privacy policies [15] for authorization to view the data, share the data, and use data for processing. The privacy consideration extends to the authentication verification process so that the authorized user can prove their identity and shall access the data legitimately. To maintain confidentiality of the data, the following considerations may be followed:

- Data disclosure policies
- Framing laws and regulations for data confidentiality
- Condition under which data can be used and disclosed
- Negative impact of data if disclosed
- Permission policies for data accessing.

The confidentiality of the data is essential for maintaining the business process continuity, and loss of confidentiality leads to the complete shutdown of the business process. The guidelines for maintaining the confidentiality of the data in the cloud network are narrated as follows:

- **Data encryption**: The sensitive data that are transmitted through the insecure cloud network must be encrypted to convert the understandable plaintext format to an unreadable format [16]. Varieties of cryptographic algorithms [17] have evolved for maintaining the secrecy of data and are discussed in the upcoming section.
- **Data access management**: The confidentiality of the data can be managed by maintaining an access policy that can access the data from the cloud infrastructure. The authorized person is allowed to access the data through successful verification of the authentication process by providing the login credential and password.
- **Security for physical infrastructure and documents**: The maintenance of data confidentiality includes controlling the access of the data from physical infrastructures. The theft of physical documents or devices also leads to loss of data. Hence, it is necessary to protect the physical devices in the infrastructure and the documents that hold sensitive data.
- **Secure disposal of device and documents**: The data in the physical devices and the documents need to be entirely recovered before disposing of them when the devices are no longer necessary or no longer provide better functionality. When sensitive data have lost its necessity, it must be disposed of properly. The sensitive data must be erased or appropriately damaged from the physical device or documents.
- **Data acquisition management**: A high level of consciousness must be followed during the acquisition of data. The privacy and confidentiality of data is to be appropriately maintained, and the data acquisition is to be performed only on suitable needs. The data collected without necessity have a probability of losing confidentiality, which in turn leads to the disturbance of business process continuity.
- **Data utilization management**: The risk of disturbing the data confidentiality can be mitigated by utilizing the sensitive data only on necessity. The mishandling of sensitive data disturbs the confidentiality and privacy of the data that are stored in the cloud network.
- **Device management**: The device management must incorporate proper security practices for maintaining the data integrity in the cloud network. The device management encompasses processes such as protecting the device from the malware attacks, and hence it is necessary to employ proper anti-virus software and whitelisting applications, installing firewalls, and employing proper cryptographic algorithms.

3.8.1 FRAMEWORK FOR TWO-FACTOR AUTHENTICATION

The privacy of the cloud resources may be enhanced by incorporating multi-factor authentication [18] mechanisms to execute a unique level of authentication. The multi-factor authentication possesses low-entropy authentication credentials in addition to the incorporation of the biometric identity of two-factor device authentication. The biometric identity may be an individual's fingerprint, voice print, digital signature, etc. The proposed framework comprises three phases, namely:

- Registration phase
- Authentication verification phase
- Transaction/communication phase.

The registration phase involves registering a new user with a unique login identity and creating low-entropy passwords along with the registration of biometric in the cloud server. This phase also supports resetting of passwords in case the user forgets the password. The submission of perfect credentials paves the way to the following authentication phase, which compares the submitted credentials with the credentials stored in the cloud database. On successful verification of authentication credentials, the secure session key is shared with the authenticated user. This is considered the third level of authentication – providing the session key. The transaction phase is activated. The secure session is established between the user and cloud resource. The flow chart for the proposed framework [19] is shown in Figure 3.6.

3.9   SECURITY SOLUTIONS FOR CLOUD COMPUTING

The responsibility of cloud security is jointly shared by the cloud service provider and the user. The cloud service provider needs to maintain the cloud infrastructure with proper security and privacy policies [20]. In turn, the user needs to access the data with appropriate credentials to make legitimate access to the cloud resources. The role of the service provider in preserving the data confidentiality is to maintain the cloud resources properly by restricting the data access, identifying the intruders by installing intruder detection [21] and avoidance systems, and configuring the network with proper firewalls such that it is possible to determine the anomalous behaviour of data packets in the network. In addition to the service provider's role, certain security practices have to be followed by the cloud users so that data confidentiality can be maintained. The role of users extends from utilizing the access privileges effectively, preserving the cloud accounts with strong credentials and passwords, employing proper cryptographic algorithms, and maintaining the security compliances properly [22].

The cloud security policies [23] are developed and employed to protect the data transferred through the private and public cloud. Cloud service providers deploy security practices and control mechanisms to preserve the cloud environments. The control mechanisms are classified into three types as follows:

- **Preventive controls**: This control mechanism is a set of processes that include enabling firewalls and anti-virus software to detect and minimize the attacks by reducing the security flaws identified by the intrusion detection system (IDS).
- **Detection controls**: The detection control mechanism identifies the attacks that are launched by the hackers. The cloud service providers need to execute security information and event management solutions for monitoring the network traffic for identifying the suspicious behaviour in the cloud infrastructure.

I

```
┌─────────────────────────┐                    ┌────────────────────────────────┐
│ Login Process: Providing │                    │ Registration of Log in          │
│      Log in Credentials  │                    │          Credentials            │
└─────────────────────────┘                    └────────────────────────────────┘
```

Login Process: Providing Log in Credentials

Registration of Log in Credentials

Credential Data base in Cloud Server

Do Credentials Match?

Login Failed

NO

YES

Second Level of Authentication using Biometric

Does Biometric Pattern Match?

Login Failed

NO

YES

Issue of Session Key to User

Does Session Key Match?

NO

Request Denied

YES

Grant of Permission of accessing Cloud Resource

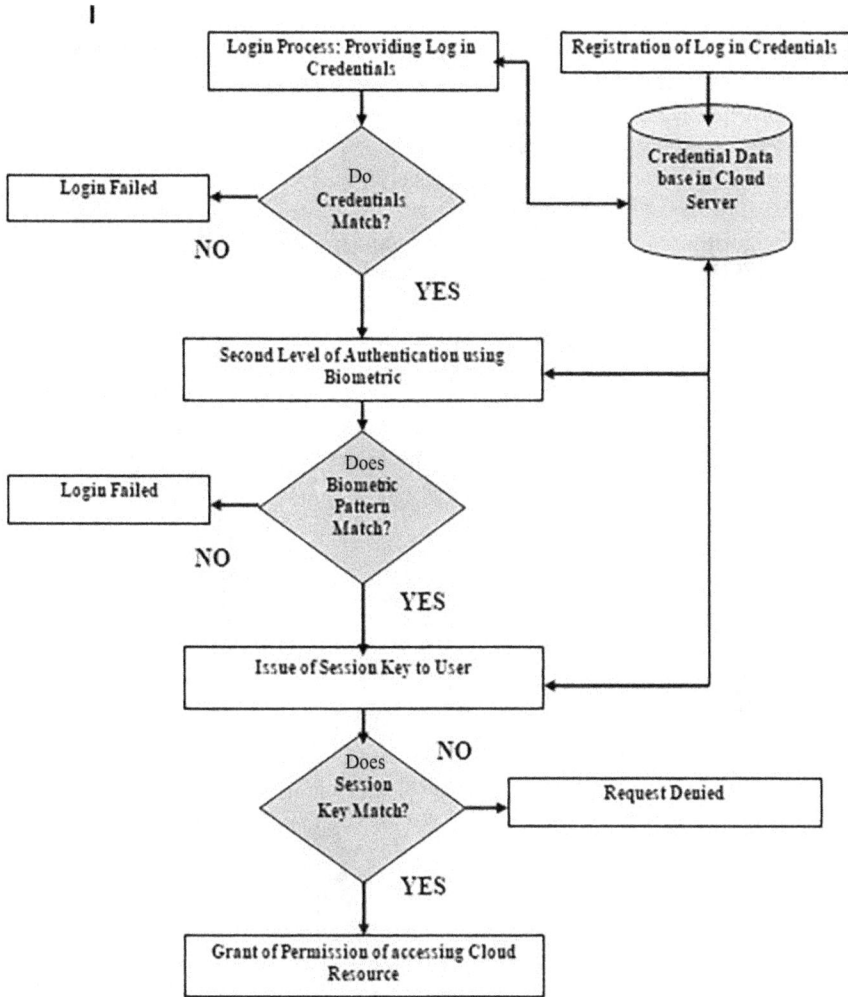

FIGURE 3.6   Flow chart for two-factor authentication in cloud computing.

- **Corrective controls**: The corrective control mechanisms are employed after the detection of suspicious activities in the cloud network. The service providers need to re-inspect the security policies of the cloud service, to take necessary backup of the data and to execute the intrusion avoidance system for detecting and removing the security bugs from the cloud infrastructure.

3.9.1   ROLE OF CRYPTOGRAPHY IN CLOUD SECURITY

Cryptography is a science of converting understandable plaintext to non-understandable ciphertext. The unauthorized users feel hard to identify the authenticated message in the cloud network. Cryptography secures the sensitive data in the cloud network, and various cryptographic algorithms evolve to offer better security preservation to the cloud data. Among the various cryptographic algorithms, Rivest–Shamir–Adleman (RSA) algorithm [24] is an asymmetric key cryptographic algorithm that employs two different keys, namely private key and public key, for encryption and decryption processes (Figure 3.7).

The RSA and other cryptographic algorithms involve modular arithmetic concept to secure the data transferred in the cloud network.

RSA is not the only cryptographic algorithm; there are also a variety of cryptographic algorithms such as data encryption standard, hash functioning – secure hash algorithm (SHA), and elliptic-curve cryptography (Table 3.1).

3.9.2 ROLE OF BLOCKCHAIN IN CLOUD SECURITY

Like conventional cryptographic algorithms, the novel blockchain algorithm [25] also plays a vital role in securing confidential data in the cloud network. The blockchain is the register for online transactions that are performed through the cloud server.

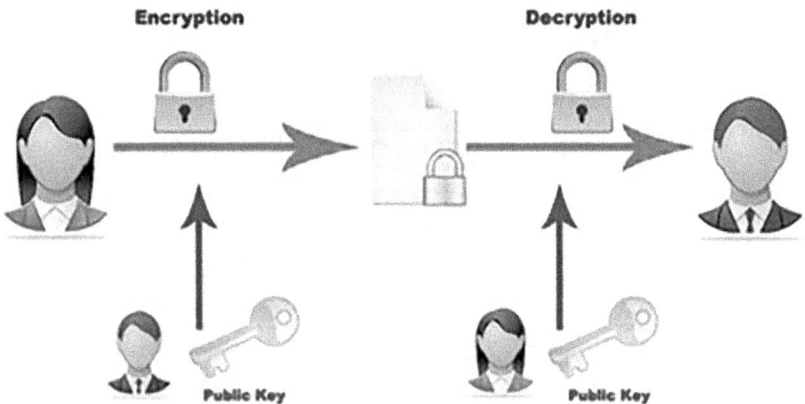

FIGURE 3.7   RSA algorithm.

TABLE 3.1

RSA Encryption and Decryption

RSA encryption:	RSA decryption:
Select prime numbers p, q	Select prime numbers p, q
Determine $n = p * q$	Determine $n = p * q$
Calculate $\emptyset(n) = (p-1)(q-1)$	Calculate $\emptyset(n) = (p-1)(q-1)$
Select integer "e"	Select integer "e"
GCD $(\emptyset(n), e) = 1$	GCD $(\emptyset(n), e) = 1$
Calculate d	Calculate d
Demod $\emptyset(n) = 1$	Demod $\emptyset(n) = 1$
Determine public key KU $\{e, n\}$	Determine public key KU $\{e, n\}$
Determine private key KR $\{d, n\}$	Determine private key KR $\{d, n\}$
$C = M^e \pmod{n}$	$M = C^d \pmod{n}$

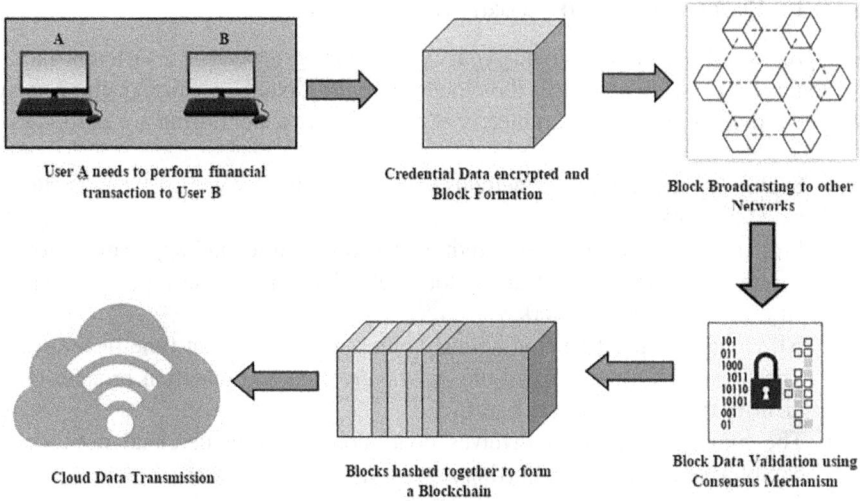

User A needs to perform financial transaction to User B

Credential Data encrypted and Block Formation

Block Broadcasting to other Networks

Cloud Data Transmission

Blocks hashed together to form a Blockchain

Block Data Validation using Consensus Mechanism

FIGURE 3.8 Blockchain in cloud security.

Blockchain technology protects data integrity and provides uninterrupted service to authorized users. Blockchain-based cloud computing supports centralized data storage, which has a threat of cyberattack, leading to the complete shutdown of the business continuity process. The blockchain and its security algorithm provide better security solutions in this regard so as to attract many users towards the utilization of cloud services with blockchain-enabled security mechanisms. Figure 3.8 depicts the role of blockchain in preserving the security and privacy policy of cloud computing. Blockchain technology allows authorized users to access the cloud server's data and provides permissions to interpret the cloud data. The advantages of blockchain in securing cloud security are as follows:

- The blockchain accesses all cloud resources that are available.
- The blockchain technology reviews the security threats of cloud computing.
- It narrates the procedure to implement the cloud operations in the desired pattern.
- It monitors the performance and the traffic of the cloud network.

3.10 SECURITY MONITORING FOR CLOUD COMPUTING

Security monitoring in cloud computing is the vital process of information storage and security management. The security principles are to be framed to provide automated solutions for the incidental concerns related to cloud computing security. The security principle must supervise the physical and virtual servers to support the continual process of data access. The benefit of security monitoring in cloud computing is that it provides simple methods to identify the abnormal behaviour of data patterns in the cloud network. The cloud security monitoring provides continual business process and incident response for the concerns that occur as a result of cyberattack.

3.10.1 PROCESS OF CLOUD SECURITY MONITORING

The process of performing cloud security monitoring depends on different methodologies and can be hosted in the cloud computing platform through third-party service providers. The vital parameters of cloud security monitoring are as follows:

- Scalability deals with the methods to monitor a huge volume of data in the cloud network.
- Proper visibility of the user activities, network traffic, and applications for continuous monitoring of anomalous behaviour and detecting a potential attack on the cloud network.
- All cloud files must be continuously monitored at constant time intervals. The files must be monitored during the creation and updation process to identify the adverse network traffic.
- The monitoring process involves continuous auditing of cloud network compliance and proper reporting regarding the compliances.

Cloud monitoring must be effective in scanning the network traffic, evaluating the traffic patterns, and identifying data before providing service to the users. The proper security monitoring enhances the network security and offers continuous business processes.

3.11   INCIDENT RESPONSE TO ATTACKS IN CLOUD COMPUTING

Normally, the security threats identified in mobile cloud computing and normal cloud computing may process some insider intrusion like a threat and the situation in the outside environment. The threat that may occur due to the insider will be arising from the intra-support networked users such as the employees and the persons who worked before and who may have had the accessibility to computer systems, including the server and the stored information. The persons from outside environment who may create threats will be from various entities. The question of integrity may be disturbed because of this threat, and it additionally includes the availability besides the confidentiality over the cloud-based resources.

3.11.1 RESPONSE TO THE THREATS

3.11.1.1   Using Firewalls

Basically, many forms of attacks can be tackled with the help of firewalls. The firewalls used in the system may provide the solution to tackle the condition of security attacks. The most important thing in using a firewall [26] is that it can indulge in protecting the computer systems network from the unauthorized users creating traffic. Considering different researches, there exist many types of firewalls. They may affect the network layer for investigating the packets information, especially the header information such as the protocol and the other types of addresses, in line with the rules defined in prior. Some of the attacks such as spoofing can be eliminated with the help of firewalls, which can be effective with the packet filtering concept.

The most effective type of inspection that may be held with the firewall is maintaining the state table by executing the connections. The controlling of the packet information is done with the help of the firewall, and it may compare the information contained in the state table to justify whether the statement is correct or not. The monitoring and controlling of the outputs and the inputs can be done with the help of the firewall, and the operation is done in the application layer. Moreover, controlling the traffic can be done with a lower layer existing in the application layer. The proxy agents used for the communication between the two hosts may dwell with the application layer. The operation of the firewalls are based on virtual machines, and it may have control over it for performing the filtering process of the packets.

3.11.1.2   Using MIDS and MIPS

Most probably, the attacks by an outsider can even be tackled with the help of firewalls. But the insider attacks cannot be traced and prevented by the firewalls. Hence, another mechanism is required for the detection and prevention of insider attacks. The usage of the MIDS (modified intrusion detection system) and the MIPS (modified intrusion prevention system) can provide a better solution for insider attacks. For intrusion detection, the dataset is gathered from the cloud server. The algorithm that supports the modified version of the intrusion detection system can monitor the various data collected from the cloud environment to detect crucial attacks. Like the MIDS, various other schemes are available to detect the attacks, such as the NIDS and WIDS. Here, the data are collected from the cloud servers and the intrusion is verified. The detection of the attacks is verified by the MIDS. The special thing in this is the ability to monitor the application data and the log files that could create suspicious attacks. The network traffic monitoring is also done. The signature- and the anomaly-based detections are mainly used for the detection of the intrusion. The comparison of the events captured and the identification of the attacks' patterns are performed feasibly for intrusion detection based on the signature. For the anomaly-related system, the system's normal activity or the network is gathered as the base data. The data or the information received or deviated from the normal process is considered the intruded events. The known attacks are detected using the signature, and the unknown attacks are detected using the anomaly.

3.12   PRIVACY PRESERVATION FOR DATA IN CLOUD COMPUTING

The machine learning approach [27] that is needed for the learning-based discussion leads to the non-trusting of the clients to secure the needs from the model used for the combined training by not explicitly exposing their datasets. The communication cost is assumed as a major challenge for considering the bandwidth of the network. Typically, the parameters used for enhancing the model are used for performing the attacks due to the inversion. To provide privacy protection to the network, a median-weighed cryptographic protocol is designed. The efficiency is enhanced by employing the perfect optimization strategy (Figure 3.9).

Cloud computing security is the most needed thing to be notified in recent scenarios. If the security is not provided to the right extent, then the data may go at high risk. This may affect the proper transmission of information throughout the network.

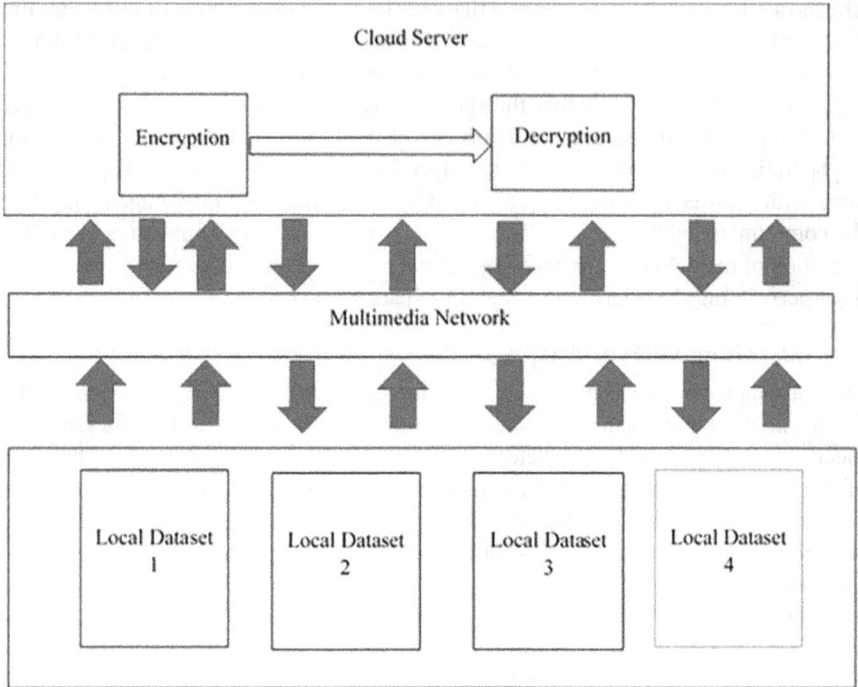

FIGURE 3.9   Privacy preservation for data in cloud computing.

Cloud computing nowadays provides a tremendous change in security-related access to the authenticated group of users. The predominant measures for the security needed to develop or identify the different security challenges have some difficulties when dealing with the already stored information of various organizations that are distributing their information over the cloud. Hence, there are severe security threats in case of misusing the datasheet. The risk generated can be eliminated, and the data are being protected [28]. But for protecting the data, some measures are to be taken.

The increase in the quality of the most relevant and important data that are being extracted by the controllers of the data is achieved by the essentials depicted in view of security alerts that are propounded in the system. The module might not be used only for storing the data and also can be used for processing the same in the premises of the cloud. Mostly the security needed for the cloud could provide the most sensitive and unauthenticated data that have some outsourced information that is essential for executing in the cloud premises. The resultant violations of the data are repeated in order to satisfy the necessity for the legal constraints for protection. A lot of researches are held to adopt the need for protection of privacy in the cloud-based server. Artificial intelligence-based techniques provide a tremendous security, focusing on cloud-based information since it needs some restoration models for data sanitation.

The artificial intelligence abilities involved in the business mode cloud computing nature will make the enterprises more productive, efficient, and understandable;

however, facilitating the information that is involved in the cloud computing environment might offer some high-adaptability deftness and cost investment funds. The two major phases of the proposed protection conservation framework are information disinfection and rebuilding. The proposed sanitization work may involve the discrepancy measure that relies upon the ideal key generation that is performed by the hybrid metaheuristic algorithm. The hybrid algorithm will merge a couple of good-performing algorithms such as modified shark smell optimization and enhanced Jaya algorithm. The combination of these two algorithms brings a novel model. The accomplishment of the optimal key generation is done by the multi-objective function deriving, which includes the parameters such as the modification degree, ratio of hiding, and information preservation ratio. Finally, the analysis of performance will prove the proposed models' efficiency in comparison with the other models reviewed by the literature survey to enhance cloud security.

3.13 ANALYSIS OF EXISTING SECURITY ALGORITHMS

The increase in the need for cloud computing leads to concerns related to security, i.e. information security, which can be overcome with the benefits of some security algorithms within the processes and the systems. Here, some of the algorithms that enhance the security of the data transmission and reception in the cloud environment are discussed. The cryptographic algorithms [18] such as the symmetric and the asymmetric algorithms are analysed and discussed in this chapter. Consider the asymmetric algorithms; they use a pair of the same keys named public key used for encryption purposes and another set of keys called private keys for the purpose of the decryption process [29]. These keys are used for the transformation of plaintext to ciphertext. Some of the most familiar algorithms used are given below.

3.13.1 RIVEST–SHAMIR–ADLEMAN (RSA) ALGORITHM

One of the most effective algorithms suited for cloud security is the RSA algorithm. It is a very efficient algorithm for data transmission. While dealing with this algorithm, the encryption process is done prior to storing the data in the cloud server. To get back the data, the cloud server needs a request from the user to enable the accessing permission. The user is initially authenticated, and the request is placed and processed. NET security framework supports the algorithm for the whole process [30]. The conversion of the data from one form to another form is made possible with a public key and private key used in the cloud service provider. For this task, this algorithm uses the concept of including the prime numbers for the generation of the private and public keys. The mathematical multiplication and the usage of formulas help to generate the keys for private and the public usage.

3.13.2 DIFFIE–HELLMAN KEY EXCHANGE (D-H) ALGORITHM

The technique used in this method will exchange the keys used for encryption and decryption. This is done by providing a shared key to exchange the information. The process is not suitable for encryption and decryption [31]. The process of ensuring the capabilities of the two parties, which do not have the needed knowledge within themselves, but they share the private keys throughout the web. The interchange of the keys is made, and the secret key is considered as the session key [32]. Hence, everyone can identify and find the third key, which cannot be found by the intruder who has both the other keys. This is most useful for preventing man-in-the-middle attacks.

3.13.3 SYMMETRIC ALGORITHMS

The symmetric algorithms can have a single secret key for sharing. This will help to encrypt and to decrypt the data and the information and is capable of dealing with large amount of data that could be taken from the cloud server. For processing the data from the cloud server, the cryptographic process is overwhelming with maximum speed. The concept of stream cipher is used for encrypting the plaintexts separately at a time [33]. The problems such as the exchange of the secret keys through the Internet, which is not secure, will cause the authenticity of the data to be a task. Here, the third person will have direct access to the secret keys and the obtained message will get decrypted. The actual fact in this is the process of having the secret key steps for dealing with the most difficult issues is generally met by the cloud servers while dealing with the massive amount of data, as the communication is made possible through the insecure Internet. The only chance is to avail a secret key that changes periodically. This kind of secure key that can periodically change will help to increase the security in case of issues related to data integrity and the non-repudiation of the messages.

3.13.4 MD5 ALGORITHM

The traditional cloud computing algorithm is MD5, which is proposed that can be used for the related transmission of information and the problems associated for the Internet of things which may process with large quality of errors with less security [34]. It provides certain advantages for the transmission of information by implementing the resources related to heterogeneous integrated network. This kind of algorithms can improve the security and thus enhances the accuracy of the information. It normally needs 128 bit of information and provides a variety of applications. It is applied for making sure that the information sent by us is right or not. It is a reliable algorithm that can provide access to the users to the maximum extent by making them enter the MD5 value for ensuring the security.

3.14 CONCLUSIONS

The data in cloud computing hold a high level of sensitivity, and they continue to thrive in the business continuity process. The security that compromises the privacy policy of cloud computing leads to disruption of service and to economic loss for the organization. The loss of an individual's data leads to exposure of sensitive data of the individual, leading to other severe criminal activities. This book chapter describes the security aspects and requirements of cloud computing technologies. This book chapter describes the cryptographic algorithms and their crucial strength in providing security to cloud resources. The above sections illustrate the different types of cloud computing attacks and the existing methodologies for securing the data stored in cloud computing. A strong but lightweight authentication system can be designed for the authentication verification process such that only legitimate users are permitted to access the data or cloud computing services.

REFERENCES

1. Joshi, J. B. D. (2004) Access-control language for multidomain environments. *IEEE Internet Computing* 8(6): 40–50.
2. Broberg, J., Venugopal, S., and Buyya, R. (2008) Market-oriented grids and utility computing: The state-of-the-art and future directions. *Journal of Grid Computing* 6(3): 255–276.
3. Armbrust, M., Fox, A., Griffith, R., Joseph, A. D., and Katz, R. (2009) Above the clouds: A berkeley view of cloud computing, UC Berkeley Reliable Adaptive Distributed Systems Laboratory White Paper.
4. Tourani, R., Misra, S., Mick, T. and Panwar, G. Security, privacy, and access control in information-centric networking: A survey. *IEEE Communications Surveys & Tutorials* 20(1): 566–600, Firstquarter 2018, doi:10.1109/COMST.2017.2749508.
5. Shyamala, R., and Prabakaran, D. (2018) A survey on security issues and solutions in virtual private network. *International Journal of Pure and Applied Mathematics* 119(14): 1183–1192.
6. Bhardwaj, A., Subrahmanyam, G. V. B., Avasthi, V., and Sastry, H. (2016) Security algorithms for cloud computing. *Procedia Computer Science* 85: 535–542.
7. Jansen, W. A., and Grance, T. (2011) Guidelines on security and privacy in public cloud computing.
8. Behl, A. (2011) Emerging security challenges in cloud computing: An insight to cloud security challenges and their mitigation. *In 2011 World Congress on Information and Communication Technologies*, pp. 217–222, IEEE, Mumbai, India.
9. Tebaa, M., El Hajji, S., and El Ghazi, A. (2012) Homomorphic encryption applied to the cloud computing security. *In Proceedings of the World Congress on Engineering* 1(2012): 4–6.
10. Ding, L., Wang, Z., Wang, X., and Wu, D. (2020) Security information transmission algorithms for IoT based on cloud computing. *Computer Communications* 155: 32–39.
11. Prabakaran, D., and Shyamala, R. (2019) A review on performance of voice feature extraction techniques, 221–231. doi: 10.1109/ICCCT2.2019.8824988.
12. Kok, S., Azween, A., and Zaman, N. (2020) Evaluation metric for crypto-ransomware detection using machine learning. *Journal of Information Security and Applications* 55, 102646. doi: 10.1016/j.jisa.2020.102646.
13. Prabakaran, D. and Sathiyapriya, H. (2019) A review on methodologies and performance analysis of device identity masking techniques. *International Journal of Scientific & Technology Research* 8(12): 1–5.
14. Belguith, S. (2015) Enhancing data security in cloud computing using a lightweight cryptographic algorithm.
15. Kim, S., Lim, S., and Kim, D. (2017) Intelligent intrusion detection system featuring a virtual fence, active intruder detection, classification, tracking, and action recognition. *Annals of Nuclear Energy* 112. doi: 10.1016/j.anucene.2017.11.026.
16. Selvi, S. A., Rajesh, R. S., and Angelina Thanga Ajisha, M. (2019) An efficient communication scheme for Wi-Li-Fi network framework. *In 2019 Third International Conference on I-SMAC (IoT in Social, Mobile, Analytics and Cloud) (I-SMAC), Palladam, India*, pp. 697–701, IEEE.

17. Tripathi, S., Bhupendra, G., and Pandian, S. (2020) An alternative practical public-key cryptosystems based on the dependent RSA discrete logarithm problems. 10.1016/j.eswa.2020.114047.
18. Wilczyński, A., and Kołodziej, J. (2019) Modelling and simulation of security-aware task scheduling in cloud computing based on blockchain technology. *Simulation Modelling Practice and Theory* 99: 102038. doi: 10.1016/j.simpat.2019.102038.
19. Zhou, Y., Guan, Y., Zhang, Z., and Li, F. (2019) Cryptographic reverse firewalls for identity-based encryption. doi: 10.1007/978-981-15-0818-9_3.
20. Weber, F., and Schütte, R. (2019) A domain-oriented analysis of the impact of machine learning: The case of retailing. *Big Data and Cognitive Computing* 3, 11. doi: 10.3390/bdcc3010011.
21. Zodpe, H., Wani, P., and Mehta, R. (2012) Design and implementation of algorithm for DES cryptanalysis, pp. 278–282. doi: 10.1109/HIS.2012.6421347.
22. Muthukumar, P. (2013) Synchronization of chaotic systems using feedback controller: An application to Diffie–Hellman key exchange protocol and ElGamal public key cryptosystem. *Journal of the Egyptian Mathematical Society* 22. doi: 10.1016/j.joems.2013.10.003.
23. Jenifer, A. L., and R. S. Kumaran (2018) Development of vulnerability scanner. *International Research Journal of Engineering and Technology* 5(7): 1278–1281.
24. Preethi, S. and Kumaran, R. S. (2016) Secure key establishment scheme for wireless sensor networks in distributed IoT applications. *International Journal of Modern Trends in Engineering and Science* 3(8): 90–93.
25. Song, H. (2020) Testing and evaluation system for cloud computing information security products. *Procedia Computer Science* 166: 84–87. doi: 10.1016/j.procs.2020.02.023.
26. Sun, P. (2020) Security and privacy protection in cloud computing: Discussions and challenges. *Journal of Network and Computer Applications* 160: 102642. doi: 10.1016/j.jnca.2020.102642.
27. Mthunzi, S., Benkhelifa, E., Bosakowski, T., Ghedira, C., and Barhamgi, M. (2019) Cloud computing security taxonomy: From an atomistic to a holistic view. *Future Generation Computer Systems* 107. doi: 10.1016/j.future.2019.11.013.
28. Subramanian, N., and Jeyaraj, A. (2018) Recent security challenges in cloud computing. *Computers & Electrical Engineering* 71: 28–42. doi: 10.1016/j.compeleceng.2018.06.006.
29. Kumar, R., Raj, H., and Perianayagam, J. (2018) Exploring data security issues and solutions in cloud computing. *Procedia Computer Science* 125: 691–697. doi: 10.1016/j.procs.2017.12.089.
30. Patil, R., Dudeja, H., and Modi, C. (2019) Designing an efficient security framework for detecting intrusions in virtual network of cloud computing. *Computers & Security* 85. doi: 10.1016/j.cose.2019.05.016.
31. Parikh, S., Dave, D., Patel, R., and Doshi, N. (2019) Security and privacy issues in cloud, fog and edge computing. *Procedia Computer Science* 160: 734–739. doi: 10.1016/j.procs.2019.11.018.
32. Jia, H., Liu, X., Di, X., Qi, H, Ligang, C., Li, J., and Yang, H. (2019) Security strategy for virtual machine allocation in cloud computing. *Procedia Computer Science* 147: 140–144. doi: 10.1016/j.procs.2019.01.204.
33. Kumar, T. A., and Rajesh, R. S. (2014, December) Towards power efficient wireless NoC router for SOC. *In 2014 International Conference on Communication and Network Technologies*, Sivakasi, India, pp. 254–259, IEEE.
34. Prabakaran, D., and Ramachandran, S. (2022). Multi-factor authentication for secured financial transactions in cloud environment. CMC-Computers, Materials & Continua 70(1): 1781–1798.

4 Networking Security Incitements in Cloud Computing

S. Maheswari
National Engineering College

A. Ramathilagam and R. Ramani
PSR Engineering College

CONTENTS

DOI: 10.1201/9781003219880-4

4.1 INTRODUCTION

Cloud computing is one of the most important advances in information technology, and it offers lot of benefits. By providing the client with the option of adopting systems with cutting-edge computing capabilities, high availability, and scalability, this technology relieves the client from the stress of operating increasingly complex and expensive systems. Security is the most important part of any kind of computing, and subsequently it's natural that security concerns are also important in the cloud. Identity management and authentication are particularly important in cloud computing, since the cloud computing strategy could be connected with users' sensitive data stored both at clients' end and on cloud servers.

Cloud computing offers an excess of benefits while, moreover, posing a number of obstacles. Security is seen as one of the most considerable obstacles within the way of cloud computing development. Cloud computing possesses lot of protection for stressful conditions that are probably dynamic and broad in every aspect. The placement of data is a critical aspect of cloud computing security. Location transparency is one of the most obvious benefits of cloud computing, but it also possesses security risk: Without knowing the precise location of data storage, data protection laws in some regions may be badly impacted and broken. Personal data security of cloud users is thus a critical problem in a cloud computing environment.

Cloud security is critical for both business and regular consumers. Everyone wants their data to be safe and secure, and the organizations have legal responsibilities to protect client data, with some industries possessing more stringent data storage regulations.

Consumers can avail profit from cloud computing systems [12] in a number of ways, including:
- Submitting an e-mail
- Data recovery, storage, and retrieval
- Designing and testing smartphone devices
- Data analysis
- Streaming audio and video
- On-demand app distribution.

Cloud security is threatened by theft [1,2], spillage, and destruction of data stored online via distributed storage stages. Firewalls, intrusion testing, jumbling, tokenization, virtual private organizations, and avoiding public web associations all contribute to cloud security. The following are the key security threats associated with cloud computing:

- Offenses against the law
- Identity theft
- Malware infections and data breaches
- Customer interest has eroded, and there can be a hazard of profit loss.

Services designed to ensure the security of network usage and data are considered as network protection, which includes software development and tools, and they can deal with various risks. They also prevent them from broadcasting the network. Network access is regulated by an effective network security. Efficient and dependable worker insurance is essential for any business because it reduces the costs of wages, salaries, liability, and other expenses that must be paid to the creditors, resulting in increased revenue and lower operating costs. Section 4.2 presents the overview of cloud computing. Section 4.3 discusses the cloud computing and security threats. Section 4.4 deals with the importance of cloud security. Section 4.5 depicts that cloud security has three key practices.

4.2 OVERVIEW OF CLOUD COMPUTING

Cloud computing may be defined as storing and getting access to information over the net as opposed to the PC's difficult force. This approach does not allow an entry to the information from both the PC difficult force and over a devoted PC community such as domestic or workplace community. Cloud computing approach information is saved at far-flung vicinity and is synchronized with different Internet information. Cloud computing refers to a computer paradigm in which a large number of structures are connected via private or public networks to provide dynamically scalable software, data, and report storage infrastructure. Since the turn of the century, the cost of computing, software hosting, content material handling, and transportation has dropped significantly. Most outstanding instances of cloud computing are Google Cloud; Microsoft 365; and administrative center via the means of OS33 which permits customers to store, get entry to, and edit their MS office files online like in browser without placing in real application in their devices.

4.2.1 Types of Cloud Computing

The garage alternative on cloud is in three forms such as public cloud, private cloud, and hybrid cloud. Installing programs on public, private, or hybrid clouds depends on the project. Cloud integrators can be crucial in deciding which cloud direction is the best for each organization.

- Third-party public clouds provide advanced economies of scale to clients by distributing infrastructure fees over a large number of users, resulting in an attractive low-cost and high-quality experience for each individual buyer. A "pay-as-you-go" version is also available. Both the clients use the same technology pool, but they have different settings, security protections, and availability. With the help of the cloud provider, these can be managed and assisted. A public cloud has the benefit of being larger than an organization's cloud, allowing scaling quickly and on demand. The major advantages and disadvantages of public cloud are listed below:

Advantages of a public cloud
- Infrastructure management is simple, and hence there is no need of a corporation to develop its own software when it is handled by a third party.
- The costs are minimal, and subsequently there is no need to invest in software or hardware because you just pay for the services that you use.
- Get ready to work 20 hours a day, 7 days a week.
- Autoscaling is a function offered by the public cloud providers. That is, all virtual machines in the public cloud system can be built, scaled, and shut down at any time. As a result, the workload can be balanced based on the requirements, avoiding downtime and crashes.
- The efficiency of public cloud services is higher.

Disadvantages of a public cloud
- Many companies are concerned about the protection and privacy of data stored in a public cloud service. Many providers offer public cloud services that are encrypted to some degree. However, the issue is with the organization and how they plan to use them. As a result, businesses must employ data protection procedures, whereas the safest option is not considered for sensitive data. Furthermore, confidence in the third-party vendor is challenged, since they may be from a different country, each with their own set of protection and privacy regulations.
- Both the customers share the same services on public clouds. Companies also have little influence and configuration over personal data, since the cloud is entirely managed and supported by the service provider.

Private clouds are tailored to the needs of a single organization, since they need to address data protection issues and possess more control than a public cloud that it typically provides. A non-public cloud comes in several flavors: on-premises private cloud. Internal clouds, also known as on-premises private clouds, are hosted in a person's own data center. This version provides a more streamlined approach and security, but it is limited with duration and scalability. IT agencies may also choose to pay for capital and maintenance expenses for physical infrastructure. Physical resources will need IT agencies to pay capital and operation fees. An externally managed private cloud that is hosted by a cloud provider offers support and a separate cloud environment with complete privacy assurance. This is ideal for businesses that are unable to use the public cloud because they still share physical resources. The major advantages and disadvantages of public cloud are explained below:

Advantages of a private cloud
- The private cloud can only be used by a single company. Hardware, data storage, and connectivity can ensure a higher level of protection.
- The private cloud is stored behind the internal firewall of the company intranet. It gives access to the same tools as the public model, but has less chance of being hacked over the Internet.

- The private cloud aims to improve the scalability of storage and computing. The resources of this platform are not shared. The internal IT team is responsible for maintaining the configuration.

Disadvantages of a private cloud
- Since private cloud requires both hardware and maintenance, it is typically more expensive than public clouds. Hardware, as well as an operating system and software application licenses, will be needed.
- It is more expensive and time-consuming to use a private cloud service than a public cloud service. It's a long-term commitment that necessitates constant attention and treatment. In-house IT management is needed for a private cloud service.
- Hybrid clouds are blending of public and private cloud models. In a hybrid cloud, service providers can leverage third-party cloud providers to entirely or partially extend computing capacity. Externally provisioned scale can be given on demand in a hybrid cloud environment. To handle any unforeseen increase in workload, the option of reinforcing a non-public cloud with public cloud resources can be used.

Advantages of a hybrid cloud
- The organization's workload can reach local computing efficiency through hybrid cloud, by ensuring effective workload management. A set of extremely scalable and versatile servers will be received, and that may be used or sold to other organizations. One of the most significant advantages of hybrid cloud is the presence of a centralized private infrastructure.
- Businesses only pay for the public cloud infrastructure portion when using the infrastructure, making hybrid cloud profitable.

Disadvantages of a hybrid cloud
- Although long-term cost reductions are done as one of the many advantages, the initial cost of creating a hybrid cloud is higher than that of establishing a public cloud. In order to deploy on-premises in a hybrid cloud environment, special hardware is required and it cuts a substantial portion of the budget.
- Hybrid clouds are actually beneficial. Expert IT professionals, on the other hand, are responsible for ensuring the highest degree of data security possible. The hybrid cloud world is inaccessible to public clouds, since certain businesses do not store data off-site.
- If cloud usability isn't carefully chosen, it can be a real pain in the neck of hybrid cloud environments. Since a quick on-premises system can't keep up with a slow public infrastructure, hybrid cloud performance becomes sluggish.

4.2.2 Cloud Computing-Based Services

4.2.2.1 Uses of Cloud-Based Services

Cloud-primarily [10,11] based totally packages are hired by groups and governments to satisfy a variety of software and era needs, along with customer relationship management, database, compute, and statistics storage. Unlike traditional IT environments, wherein departments finance software program and hardware as well as installation over months, the cloud-primarily based totally structures provide IT sources from minutes to hours as well as match the prices for the usage actually. As a result, the agencies that have extra flexibility are better equipped to manipulate their spending. Meanwhile, the customers can get entry into cloud-primarily based totally services from any web-linked tool to streamline software use; store, distribute, and guard information; and get entry to content material from any web-linked computer.

4.2.2.2 Common Attributes for Cloud-Based Services

- Make a new app with new features and services.
- Data can be stored, backed up, and recovered.
- Make websites and blogs and host them.
- Streaming of audio and video is possible.
- On-the-go app delivery.
- To delegate and reallocate resources without problems, cloud computing makes big use of server and garage virtualization.
- In lots of networked computers, gears are accessed via an Internet browser or a skinny client such as computer, smartphone, and so on.
- Resources may be scaled up or down automatically.

4.2.3 CLOUD COMPUTING SERVICES WITH A VARIETY OF SHAPES AND SIZES

The services of cloud providers can be classified into three groups, as indicated in Figure 4.1. Model of software-as-a-service: Software-as-a-service (SaaS) provider, with the preference of being established and maintained on consumer machines. The service is generally to have on a month-to-month or annual subscription foundation, and it may be accessed through the Internet.

The rise of fully cloud-based computing correlates with the growth of software-as-a-service (SaaS). Business people, who have desired to improve their packages, need to buy compact disks with updates and import them onto their computer systems earlier than the usage of SaaS. Software is managed by computer systems that are owned and operated by the government. When using SaaS, users do not need to upgrade any software. Customers can instead access the service by connecting to the service provider's network via the Internet using a web browser. SaaS can be used for a variety of items, such as services for e-mail, functions in auditing, and automating product and service sign-ups. Document management, including file sharing and collaboration as well as calendars shared by the whole organization, can be used to plan activities.

Platform-as-a-service model: The PaaS issuer presents all software programs and hardware to broaden and run cloud-primarily based totally packages over the general public Internet with a dedicated community link. PaaS provides developers with a framework from which they can design applications. PaaS enables quick, convenient, and cost-effective application development, testing, and deployment. Using this technology, corporation operations or a third-party provider can cope with

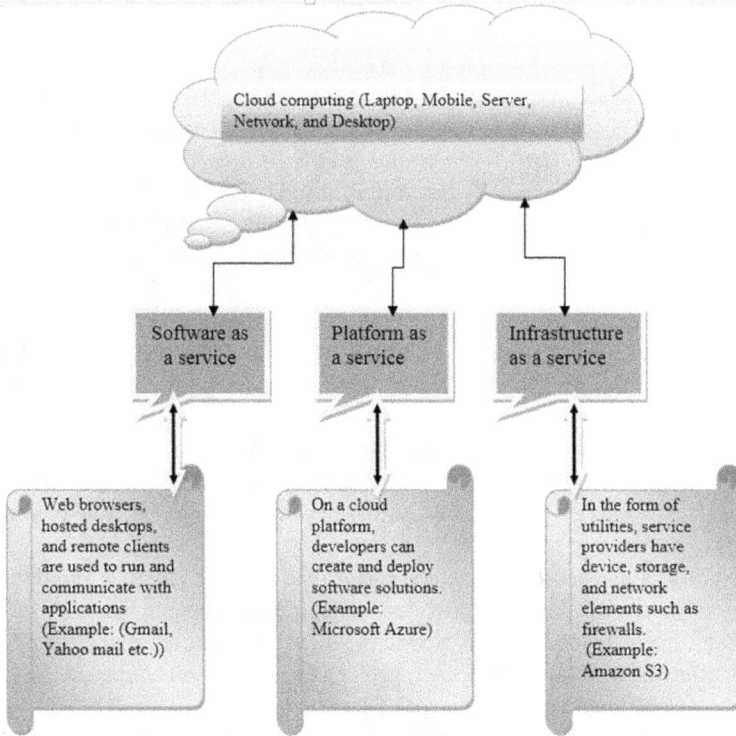

FIGURE 4.1 Cloud computing-based services.

working systems, virtualization, servers, storage, networking, and the PaaS application. The programs, on the other hand, are managed by the developers.

PaaS can be used for a variety of items, such as creation and management of application programming interfaces (APIs). APIs and micro-services can be created, run, managed, and secured using PaaS. Control of business processes and database management systems are also involved among other things.

Infrastructure-as-a-service model: Infrastructure-as-a-service is a famous cloud service transport version wherein customers pay a third-party cloud service vendor to get entry to the controlled IT infrastructure. In an IaaS model, a cloud provider hosts infrastructure components such as servers, storage, and networking equipment, as well as virtualized host machines, which are typically located in an on-premises data center. Billing, security service, scrutinizing, clustering the data, and replication of these data are all examples of the infrastructure services provided.

IaaS can be used for a variety of items, such as cloud-based software testing and development. IaaS infrastructure is the best for hosting and supporting web-based applications, and it has the potential to scale up resources to support the application during high-demand periods. Cloud IaaS is suitable for big data analysis. Amazon Web Services is a cloud computing IaaS service. Google Compute Engine is a computer program that runs on Google.

4.2.4 REAL-TIME EXAMPLES OF CLOUD COMPUTING

Cloud storage is ubiquitous, and it is on a daily basis across a variety of platforms. Cloud computing has an effect on the lives of human in a variety of ways as expressed below. Google's Gmail is one of the best instances of cloud computing.

- Zoom is a cloud-primarily based totally audio and video conferencing software program utility with the facts of conferences, and it saves them to the cloud so that customers can get right of entry to them from everywhere at any time.
- Dropbox promotes teamwork via the means of permitting customers to proportionate big files. It has a massive PowerPoint, Photoshop, or Sketch challenge that desires the team's assist however is clogging up your inbox? It's on Dropbox.
- Without provisioning or maintaining servers, AWS allows the developers to run code for any application or backend service. The pay-as-you-go model scales up and down with a company's needs, allowing real-time changes in data storage and usage.
- Netflix is an instance of an enterprise that uses the cloud. It provides billions of hours through the provider to clients everywhere in the globe through working on Amazon Web Services. Users can also additionally order its products/offerings through the usage of PCs, tablets, or cellular gadgets from nearly everywhere within the globe.
- Google Drive is a natural cloud computing service that consists of all online garages, and it may be used with the subsequent cloud productiveness apps such as Google Docs, Sheets, and Slides, which are all unfastened to use.
- The following three are cloud-primarily based totally natural language clever bots: Siri, Alexa, and Google Assistant. These chatbots employ cloud's computing electricity to offer personalized, context-applicable patron interactions. Recollect that the following time you say, "Welcome Siri!" there's a cloud-based AI-driven response at work.
- The bulk of messaging and calling apps, such as Skype and WhatsApp, rely on cloud infrastructure. Rather than being saved on the mobile PC, all your messages and information are saved on the service provider's hardware. This allows to access the information from anywhere in the Internet.
- A cloud provider issuer is likewise used for lots of enterprise control systems, inclusive of patron courting control and employer aid planning. Software as a service has grown with reputation as a way of turning the employer-degree software. Popular examples of this version encompass Salesforce, HubSpot, and Marketo. Both the provider company and the purchasers have gained from this technique, due to its cost-efficiency and reliability. It permits to control, maintain, and stable the organization's critical enterprise assets with ease, and it permits to get admission to those packages through an Internet browser.
- The cloud will help to save money while also allowing to obtain the apps to the market faster. Developers can use the cloud to create and dismantle test and development environments instead of building up physical

environments. This makes it difficult for the technical team to get resources and devote time and effort to the project. These development environments may also be scaled up or down depending on the needs. Common testing tools include LoadStorm and BlazeMeter.

- Cloud systems have proven to be a viable solution for developing online, smartphone, and even game applications. Adaptive cross-platform experiences can be easily developed for the users by using the cloud. Directory services, search, and security are among the pre-programmed tools, and libraries are also included with these platforms. Hence, the development process will be streamlined and accelerated. Amazon Lumberyard is a popular cloud-based platform for developing mobile games.
- Cloud services allow large-scale data mining by supplying more computing power and sophisticated software. Parallel computing, Cassandra, and high-performance cloud computing are just few examples of open-source cloud-based big data solutions. It would be exceedingly difficult to collect and analyze data in real time without the cloud, for small businesses, in particular.

4.3 CLOUD COMPUTING AND SECURITY THREATS

4.3.1 UNDERSTANDING ABOUT CLOUD SECURITY AND ITS BENEFITS

Cloud data security is essential, once to make sure that the stored data are safe in cloud and security is extremely much significant and top priority for all quite cloud storage services. Cloud security [3,4] is also referred to as cloud computing security, which comprises a bundle of strategies, authority, method statements, and applied science that are effective when collectively in the direction of guard cloud computing, various informative data, and engineering system features. Those well-being highlights have been arranged toward safeguarding cloud information, directing administrative consistence, and protecting customer's privateness, moreover, as placing verification rules for clients and gadgets. Cloud security shall be framed for the precise requirements of the business from verifying access to clear traffic because these rules shall be structured and handled well in a location credits to which management indirect cost are decreased and information technology teams shall be authorized to focus in new area of business.

Cloud security [7] is important for both industry and individual users. Every person desires to understand whether their data are safe and secure. Most of the businesses have reciprocal non-disclosures, non-circumvention, and accuracy agreement to take care of client data in a secured manner, and certain business sectors/verticals follow more non-compromise rules about data storage.

All crucial components of cloud security are purposefully designed for maintaining information and system compliance reports such as purchase orders, job cost reports, performance reports, confidential documents (patent rights, etc.), and financial analysis. Avoiding data leaks and theft is vital for upholding the customer's mutual trust and for shielding the assets that contribute to competitive advantage.

Most of the businesses have been shifting their existing applications and data to the cloud platform, and this trend is increased consistently in recent days. The cloud platforms afford all range of firms with an efficient secure outlet for storage of relevant and different kinds of information. But no firm/organization can delineate the advantages of cloud computing without the guarantee of safety for its end user programs. Why cloud security is crucial for every enterprise? Top reason for critical-ity of cloud security for each and every company is high-risk data. Presently, cloud security is an all-time rapid-growing concern, despite the varied set of merits that it provides to the individuals and corporations alike. Based on recent analysis, 90% of firms remain unconvinced about trusting 100% on cloud infrastructure merits for the cloud security issues.

Even though there are numerous issues on 100% trust on cloud, there are tons of things that the organizations are yet to understand, when it involves cloud security. It should always be understood that the cloud isn't an insecure platform and it's the safety model which is comparatively different from other platforms obtainable in the market at present. Moreover, there are tons of latest relationships to be fulfilled with reference to data storage.

Cloud center shall be quite secure, particularly safer than their traditional coun-terparts. The purpose of maintaining a robust cloud security posture will help the organizations to realize the following benefits:

- Reduction in upfront charges
- Reduction in operational expenses
- Ease of business volume
- Increasing consistency and accessibility
- Establishing new approach of functioning.

Centralized security: Normally, cloud security [8] centralizes knowledge shield. Cloud-based networks contain multiple devices and end nodes that can't be easy to tackle, particularly handling shadow IT or BYOD. Management of those entities cen-trally increases traffic analysis and web filtering as well as refines the network trial and outcomes with lesser software and strategy updates. Disaster recovery strategies shall be performed and executed without difficulties once that are tackled in one location.

Reduced costs: Utilization of cloud storage and security is to eliminate the necessity to obtain a place in devoted hardware. Additionally, it reduces expenses. Somewhere once the information technology teams work on security problems immediately, cloud security will suggest proactive security measures that provide safety 24/7 with tiny or no human interference.

Reduced administration: At the time of employing a reputable cloud security platform, manual security configurations shall be avoided and it is urged to use con-stant security updates. Cloud security platform enables the entire security adminis-tration to happen in one location, and it is completely dealt with automatically.

Reliability: Cloud computing platform provides the last word with reliability. With perfect cloud security measures, clients can securely access various information and software within the cloud regardless of user location and usage of type devices.

Different kinds of cloud models are inclined to various sorts of threats. Information technology departments are naturally precautious about moving mission-critical methods to the cloud and vital safety measures are placed suitably, while running a local, public, private, or hybrid cloud. Security of cloud recommends the entire functionality of established IT safety measures and permits the industries to exploit various merits of cloud computing as well as ensures that data confidentiality and system compliance need be adhered to.

By using cloud storage platform, data are protected in the cloud instead of being stored on-site or in close proximity. Some companies are still backing up to tapes, or storage of various information backups on-site or at an in-depth off-site position. If there is an area disaster, then that might end in both backups being lost. But cloud security platform nullifies that problem completely and protects ongoing business from the threat of knowledge loss.

Cloud data security becomes vital, as business team moves their devices, data centers, and business processes to the cloud platform. Guaranteeing and accomplishing quality cloud data through in-depth security arrangements, authoritative traditions of safety, and cloud security measures. The selection of exact cloud security solution for the dealing is extremely important to urge the simplest form of cloud, and it can guarantee that the organization is shielded from illegal access, data breaches, and other threats.

4.3.2 Open Challenges in Cloud Networking Security

The purpose of the cloud [6] requires a network that is more flexible. Running one or multiple of these virtual systems definitely requires the network that is to be restructured within milliseconds. Networks can be allowed in cloud environment profit margin entirely. These visualized cloud networking concepts, which include on-demand network assets in such a time period, are companionable with the distribution of worked out resources in a cloud currently. Cloud networking suggests security, privacy and reliability, authentication, endorsement, and appraisal mechanisms. It demonstrates the research disputes of providing protected cloud network. Cloud computing [8] platforms are possible to suffer from various recognized vulnerabilities, facilitating attackers to acquire free computing services, lift information from cloud clients, and enter the environment remaining in user location through cloud connections.

Virtual networking security: By allowing communication between different virtual components in cloud computing, virtual networking possesses novel security challenges. With a virtual network client's perception, the system could be private or confidential, while the correspondence takes place through a public infrastructure in practice. Hence, the mechanism to shield this correspondence requires to be customized. An alternative process is to perform it in every virtual device, and the same suggests that the virtual network clients have to look after communication security. As an additional substitute, the communication security is demonstrated by the virtual network supplier as a service, which recommends that the communication is properly protected by default and crystal clear for the client. By visualizing networks and their components, attacks arise; it should be identified, due to the abstraction layer commenced by means of virtualization. Hands-on practice/method has to be completed to adopt this new form. Particularly, for the reliability of virtual topology and its devices, the security routing of these networks has to be dealt with. In addition, the disputes are correlated with the existing cloud computing in the virtual networks. It comprises the following queries:

- On what basis the virtual network supplier guarantees a particular network capacity to a customer?
- What is the authentication to access the current virtual network?
- What way the virtual network consumption is accounted for?

Cloud networking of secure management: For the organization of cloud networking, accesses to the physical structure and to the network stuffs are necessary. This security admission have to be performed as one interface, where a client can state some factors on demand. According to that shared access with the physical virtualization environment, new network outbreaks arise in the network environment. One dispute is to describe the rules for entering the management interface as well as the way to execute these rules. In addition, the strategies for running virtual environments in space should be dispersed. Those strategies may describe a place where a virtual environment is authorized to run, because physical environment location finds out the authorized limits that pertain to the virtual environment.

4.3.3 SOLUTIONS FOR CLOUD SECURITY

Companies looking solutions for cloud security must think the subsequent benchmarks to unravel the challenges of visibility and control on top of cloud data as shown in Figure 4.2.

- **Cloud data visibility**: An entire observation of cloud data needs to admit the cloud service directly. The keys of cloud security achieve this by using an API link with the cloud service. By this API link, it is expected to analyze:
  - What type of data is warehoused in the cloud?
  - Which client can utilize cloud data?
  - The positions of consumers of cloud service access.
  - Who can share data with?
  - Where is cloud information found?
  - Which device is being used to access as well as fetch the cloud data?
- **Control on top of cloud data:** When you have got visibility toward cloud data, use the controls that aptly fit the industry. Those controls consist of:
  - **Classification of data**: Classify data on many stages, such as accessible, synchronized, private, or public, because this is made within the cloud. On one occasion, if the classification is done, the data are stopped from inflowing or exiting the cloud service.
  - **Data loss prevention (DLP):** Apply a cloud DLP solution to guard various information from illegal admission and routinely stop admission as well as transfer of knowledge once distrustful movement has been noticed.

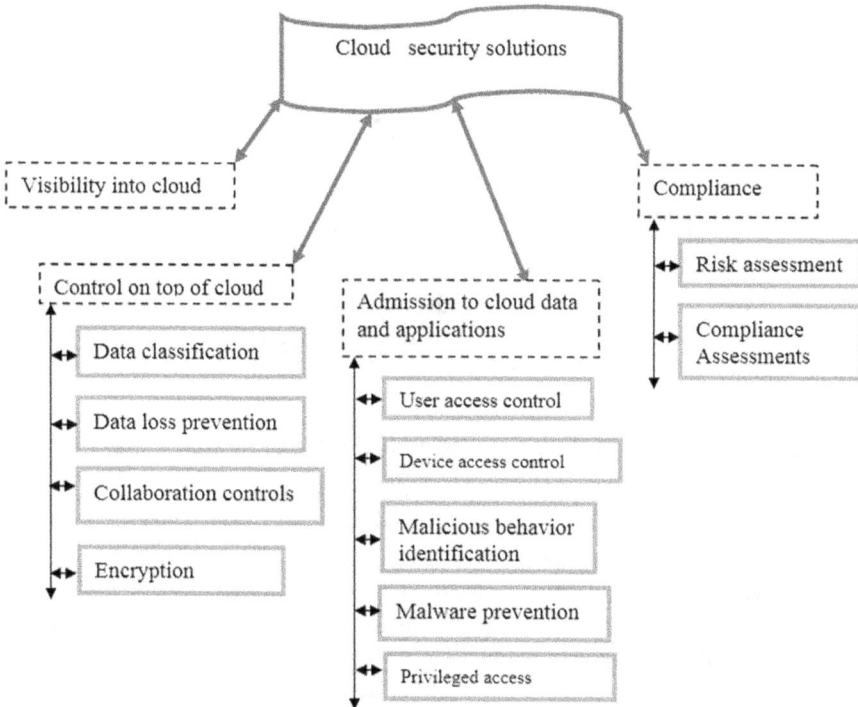

FIGURE 4.2 Cloud security solutions.

- **Collaboration controls:** Hold controls in the cloud service; for example, decline file and directory authorization in support of specific client viewer by eliminating approval, as well as cancelling shared links.
- **Encryption:** Cloud encoding frequently won't prevent illegal access to different kinds of information, even though these data have been removed.
- **Access to cloud data and applications:** Like internal security, acquire control may be an important element for cloud security. The characteristics of controls consist of:
  - **User access control:** Execute system and program acquire controls that guarantee just the allowed user's access to cloud information and software. A cloud access security broker (CASB) frequently won't enforce admit authentication.
  - **Device access control:** Deny right of entry when a private, not permitted physical/virtual machine aims to acquire cloud information.
  - **Malicious behavior identification:** Identify cooperated accounts and insider intimidations with user behavior analytics (UBA), since the malicious data exfiltration doesn't take place.
  - **Malware prevention:** Avert malware from incoming cloud facilities through methods such as port scanning, cloud-centered malware identification, and Internet traffic investigation.

- **Privileged access:** Find out complete expected kinds of admissions that the confidential accounts may need the information and programs, and locate at-site controls to alleviate experience.
- **Compliance:** Present compliance needs and practice must be improved to incorporate information and programs that exist within the cloud.
  - **Risk assessment:** Evaluate in addition to renew risk assessments to contain cloud services. Recognize as well as deal with risk issues commenced by cloud platforms and provider. Risk records for cloud providers exist to speed up the evaluation procedure.
  - **Compliance assessments:** Evaluate in addition to renew compliance estimation for HIPAA, PCI, Sarbanes–Oxley, and other relevant regulatory needs.

4.4 TECHNIQUES AND SIGNIFICANCE OF CLOUD SECURITY

According to the direction of existing exploration, one out of four concerns utilizing public cloud services experiences proficient data robbery by a harsh performer. Further, one out of five has practiced an innovative violence in contrast to their public cloud organization. In that study, 83% of establishments have specified that they collect profound facts in the cloud. With 97% of establishments all-inclusive using cloud amenities nowadays, the aforementioned issue is necessary that everybody assesses their cloud safety and improves a way to save their information from harm. Cloud retreat beginners

- Authentication
- System protection
- Investigation
- Data integrity
- Cloud security organization
- System security
- Uniqueness and access management
- Exposure and extortions.

Cloud computing being a recent skill offers many rewards. So on hitch of those profits, one possesses toward thorough study as per various cloud retreat methods as conceivable. Vulnerabilities, malicious code intrusion, hijacked accounts, and full-scale data splits are all possible issues. Supported collected works pursuits and analysis efforts, variety of the most cloud peculiar liabilities and dangers are recognized which one essential contemplate earlier creating decision to migrate in the direction of cloud for picking the facilities are shown below:

- Data breaks/data defeat
- Rejection of service attacks/keylogger inoculation
- Snatching integration

- Insufficient modification mechanism and misconfiguration
- Timid boundaries and unfortunate API executions
- Insider coercion
- Inadequate authorizations and uniqueness/conceded financial statement
- Weak control plane/inadequate conscientiousness
- Communal liabilities
- Immoral usage or exploitation of cloud services
- Deficiency of cloud security policy/regulatory violations
- Restricted cloud usage reflectiveness.

4.4.1 DATA BREAKS/DATA DEFEAT

Cloud computing, as well as service existence, is somewhat different and supports gain access to remote data through the web exists that the utmost susceptible basis for misconfiguration otherwise corruption. These actual fundamental things of cloud develop exclusive set of features that make more susceptible to complete sorts of statistics gaps. Data openings or damages are often considered as some sort of cyber safety violence; otherwise, complex info is whipped, observed, or employed via an illegal alien or it gets the outcome available for unintended removal by service supplier or a regular calamity, such as fire burst or underground eruption. This might result in the loss of possessions (IP) to opponents and the effects of reasonable limits, economic sufferers out of monitoring suggestions, disturbing trademark price and concern organization as well as overall market price could also remain on stake because of this one adoptive caution from the customers and the commercial associates. However, encryption methods can keep records nevertheless on the value of system execution. Hence, strong and well-verified data break expectation, data damage avoidances, data tie-up and retrieval, and data supervision approach must be implemented in advance by building up awareness to drift near cloud.

4.4.2 REJECTION OF SERVICE ATTACKS/MALWARE INJECTION

The simple structure of cloud that deals with scalability as well as rapidity also develops rear ground for distributing great accessible keylogger. Cloud requests themselves remain great big stick used for scattering the mischievous outbreaks on an outsized scale to originate better destruction similar to stealing accounts and breaking records. Malware insertions are mostly system draft which is inserted into the essential cloud tune-up unit and hence, it follows because valid case encompass access to all or any of the complex resources other than detail intruder know how to snoop, cooperate the general reliability of important information. Denial of service (DoS) attack builds treasured services which are engaged for the legal customer, and hence, it hinders the general concert and protection. DoS acts as a hacker, and it is used like a model for hiding malicious behavior behind firewalls by allowing it to spread more quickly and cause more damage than malware.

4.4.3 SNATCHING INTEGRATION

The latest development of straightforward variation of cloud services through organization results in different sets of problems associated with capturing report. Imposter nowadays can simply abuse the facility of getting contact with login permits, and consequently, the susceptible data consist of trade logic, tasks, data, and requests kept on the isolated cloud. Account capture that comprises reused password, scripting bugs, cross-site scripting permits the invader to fake and deploy information. Phishing, man-in-cloud attack, keylogging, and buffer inundation are further threats that ultimately lead to stealing of consumer token which the cloud platform utilizes to authenticate each individual without the involvement of login permit normally in data updation or synchronization. The effects of account stealing are often cruel, and several constants result in considerable interruption of business actions by means of entire eradications of possessions and skills. Hence, account borrowing needs to be dealt with fatally as noticeable and impalpable effect out of leakage of susceptible as well as special data possibly will break the status and crew significance.

4.4.4 INSUFFICIENT CHANGE CONTROL AND MISCONFIGURATION

Capacity, as well as the scale of varied possessions utilized in cloud surroundings, has been increased by convolution plus vitality of assets with foremost dispute in configuring when it is proficiently used for professional exercise. Improperly organized valuable computing property directs to make these resources to yield intention for susceptible spiteful unsought activities, and therefore, the whole cloud repositories may possibly uncovered to trespasser. The general industry collision lies depending on the character of misconfiguration and the way rapidly it's been perceived as well as determined. Unnecessary detrimental authorization, clear admittance to ports and facilities, unsecured record storage, unmovable evasion identification and design settings, hindering ordinary protection controls, and logging and screening are few distinctive concerns related to misconfiguration, and they should be addressed with maximum concern by constantly checking for misconfigured means in valid point as conventional change control. Further, the configuration management procedure develops in an unproductive way within the cloud environment.

4.4.5 TIMID BOUNDARIES AND UNFORTUNATE API EXECUTIONS

APIs exist a boundary by linking the system and out of doors un-trusted article mainly uncovered parts of a system available through the web, make the clients easily to modify their cloud knowledge as well as ultimately afford the protected tool or ingress point for outsiders. A poorly intended weedy set of interface depicts society-valuable susceptible resources to varied protection problems associated with privacy, reliability, accessibility, and liability. After giving the programmers the tools to create and amalgamate their requests by means of additional job-critical software, API similarly serves up to verify, give rights to use, and achieve encryption. The cloud possessions are often cooperating, but the susceptibility of an API which deceits within the communication that occurs among applications is broken. Hence, typical and untie API structures have to be denoted while crafting the boundaries which can aid to guard beside both unintended and malicious endeavors to bypass the security.

4.4.6 Abettor Coercion

The individual intrusions in data security have numerous features and lots of basis. The abettor individual part could also be from several orders; both service supplier and consumer establishments can neglect their approved accesses to the establishment's or cloud provider's systems, networks, and data, as they're individually placed to create damage deprived of even breach the firewalls and further protection security reaction. There are many aspects of human security to education, and by agreeing to and operating at a level of trust, these users can use data in improper or malicious, or careless ways. Or they can perform destructive activity via malware. A variety of trials to alleviate the results of internal threats comprises routine checks of on-site/off-site servers, recurrent alteration in passwords, limited confidential admittance to security arrangements, and central servers to restricted numbers of workforces aside from monitoring access as well as proposing business collaborations to the workers. As a preventive measure, handling such types of threat might become costlier and sophisticated because it occupies restraints, forensic analysis, acceleration, observation, and checking.

4.4.7 Inadequate Identification and Uniqueness/ Compromised Financial Statement

It can aim the whole resources of cloud alongside the user establishment's assets and even affect the establishment's executive consumer also. Other residents of an equivalent cloud are moreover at great risk to security occurrence and openings. An automatic standard revolution of cryptographic passwords and keys, elimination of idle credentials, execution of right scalable central programmatic credential administration system, and usage of multifactor validation process are some of the measures that should be accepted by the cloud supplier to depart the hazard of knowledge break. Furthermore, an appropriate diligence must be considered to make sure that the third parties to whom cloud supplier may have outsourced the operations or continuation work satisfy the necessities of security as agreed by cloud service supplier because it ultimately increases the dangers and compromises the general security. Sternest credential admittance, multifactor validation, and isolation of separated accounts are some of the recommended measures that one must choose to alleviate the danger, and this could result in illegal access to data as well as knowledge. As a result, cruel intruders masked like authentic users can influence the susceptible data. Insufficient record, individuality, or data susceptible to input management are dealt with. But the fake directs to realize the admittance of cloud user's credentials.

4.4.8 Weak Control Plane/Lack of Due Conscientiousness

Irregular data arrangements, unusual APIs, and undue dependence on cloud supplier's exclusive tools formulate complex and exclusive affairs to drift from one seller to other. This might fallout in which cloud supplier will create or develop or simply just in case if for a couple of reasons that the cloud supplier stops its process and drives out of business by working data to other in suitable manners. Subsequently, it becomes frantic and ultimately may end in loss of knowledge, too. Consequently, to evade such severe conditions of seller lock-in, ample control plan and due carefulness must be in place before creating the conclusion of transferring to any cloud. Any speedy judgment exclusive of antedating the standard and nature of facilities from cloud supplier may constitute security risk, particularly when the stated services are assured to be under control by legal and official responsibilities or services are borrowed for handling subtle or private or financial data. Cloud service users have to accomplish due meticulousness and make sure that suggested cloud service supplier possesses a sufficiently strong control plane in order and lack of this might lead to loss of data by either stealing or corruption. In addition to the practical problems conferred above, one alike significant limit that must tend due weightage in deciding procedure, is individual's factor. If a private responsible is incapable of applying total control in excess of infrastructure, data security, and authentication, then integrity, security, and stability of data could even be in hazard.

4.4.9 Common Vulnerabilities

Multiple tenancy characteristic of cloud creates cloud facilities cost-efficient for individual party, but by the way, it results in one more protection problem. Development of system and software susceptibilities inside cloud setup services result in crash to retain physical as well as logical partition among diverse renters in multi-tenant situation. This malfunction to take care of partition can additionally be broken by invaders to realize illegal admittance from one tenant's source to others'. These kinds of attacks can be accomplished by abusing the susceptibilities of either cloud supplier or any other renters whose protection is more susceptible. This may lead to escalating the attack surface, resulting in an improved option of knowledge outflow. Additionally, the cloud security by evasion is a united liability of cloud service supplier and client establishment, and hence, suitable perceptive is important to implement actual security. Letdown to realize this faultless assimilation for security execution may outcome in data and sources being conceded.

4.4.10 Immoral Use or Exploitation of Cloud Services

Trespasser by abusing the susceptibilities of cloud computing resources could target consumer's or cloud supplier's assets to crowd malware tricks. A trespasser can launch a DoS attack and make the service uninvited. Alternatively, these resources are being used for criminal purposes, such as cryptocurrency mining, direct click routes, violent attacks on security breaches, and temporary trusts. The schedule might exist significantly high as activities similar to removal need massive resources. Invaders may utilize the clouds' incomparable storage size to accumulate and disseminate malware and unlawful activities resembling by giving out of reproduced books, videos, software, or music as well as invite authorized concerns in intellectual copyright fines and arrangements which may be even more expensive. In addition, the complexity of cloud deployment services causes hackers to write over time, and those unintended weak threats, risks, and vulnerabilities have further become regular with service providers and legitimate users. To control the immoral usage and exploitation of cloud services as well as to alleviate the risks possessed by cloud service use, one must need to obtain security methodology for vigorously observing cloud infrastructure consumption and develop appropriate security procedures which should classify what are the valid and suitable performances and what are the results of cruelty and techniques of spotting such behaviors.

4.4.11 Deficiency of Cloud Security Strategy/Regulatory Violations

It is imperious to create a robust cloud security policy, rules and hazard management strategy be supposed to work out before creating mind to drifting to cloud supplier for several services rather than merely lift and modify with none owing assiduousness. Mostly, several organizations are hurdled, as a result of force to suit certain guidelines, regulations, and law of land of cause, and these agreements must be the central point for complete security policy. Important health information, personal student data, personal economic information, proprietary data, research data, and personal business strategies are typically the types of knowledge that many services move to the cloud for data protection. Many are subjected to the permanent body or committee. Further, any kind of breach leads to severe punishment and judgment. Security technologies and systems need to be tailored to the goals and objectives of the business.

The underlying business priorities and the strategies must be aligned with security architectures and frameworks. Cloud supplier being a third party, ahead of approving to supply the services, also turns responsible for broadening the acceptable security trials, as weak security can cause failure, reputational damage, official consequences, and fines.

4.4.12 Inadequate Cloud Handling Visibility

The instant establishment that makes a decision to wander the possessions and actions to the cloud creates down the general discernibility and manages those assets. The power to form a choice, envisage, and evaluate whether the services accessible by clouds are secure or cruel chooses the level of visibility of cloud consumption. Albeit the establishment has selected the services of cloud supplier, it's necessary on their portion to achieve investigation and runtime checking. To reinforce the cloud visibility with less danger, it's essential to extend wide-ranging solution that brings people, development, and skill at a common place and expound recognized cloud practice strategies to all stakeholders. If not, ignorance of the organization's policies and management principles can go too far in accessing public data and putting cloud containers at risk so that services cannot be coordinated. As a result, lack of leadership, anxiety, and lack of civilization possess a high risk. Wall installation, implementation of extensive and unreliable processes, timely investigation of external activities, and defect management can help to prevent alleged behavior and completely mitigate risk.

4.4.12.1 Network (Traffic Scrutiny and Implicit Patching)

As a vital part of the safety problem, network traffic scrutiny is often the road to protection besides zero-day attacks and it makes use of known vulnerabilities. Furthermore, it may offer security by virtual reinforcing. A firewall within the cloud is faintly dissimilar from a standard firewall, since the most completing dare has the ability to organize the firewall during a way that it doesn't interrupt network links or accessible applications, during a practical private cloud or a cloud network.

4.4.12.2 Cloud Illustration (Load Security at Runtime)

Protection language and standards alter to serve the accepting mechanism that needs to be secured. Within the cloud, the perception of workload may be a unit of competency or quantity of effort that's wiped out at a cloud occurrence. Keeping workloads in contrast to exploits, malware, and illegal alterations may be a dispute for cloud administrator, as they track in server, cloud, or container surroundings. Workloads are excited up as required vigorously, but each request should be visible to the cloud manager and be ruled by means of a security strategy.

4.4.12.3 DevOps (Container Security)

The software component in cloud services has become essential in recent years. Using containers makes sure that the software can run constantly well without worrying about the particular computing background, which may turn into complex to duplicate, for example, certain set of laws, system libraries, tools, or maybe software versions. Cluster security, say for consumers of Kubernetes, also shouldn't be ignored. For creators and process teams particularly, the addition of security during software improvement becomes more applicable, as cloud-first app creation becomes general. This depicts that the containers should be perused for malware, susceptibilities (even in software dependencies), secrets or keys, and even compliance violations. Those security verifications are done during the build, if possible within the continuous integration and continuous deployment (CI/CD) workflow.

4.5   CLOUD SECURITY HAS THREE KEY PRACTICES

- **Gain visibility**
  It ensures that you have a window into the workload, equally in real time and historically.
- **Trust but verify**
  Confidence is essential to move and build quickly. But validating business-critical activity is a key for managing risk effectively.
- **Monitor and investigate**
  It uses policy-based behavioral checking and investigates with cloud perspective. In a variety of ways, cloud computing has its individual collection of benefits. However, keep in mind that the protection guarantees are not assured; however, they are possible. As mentioned above, taking some safety measures will go an extended way to trust your files are safe inside and out of the cloud. Figure 4.3 shows the security of different kinds of clouds.

4.6   CONCLUSIONS

A few security approaches, overview of cloud computing, cloud computing services, cloud security solutions, open challenges in cloud network security, and techniques

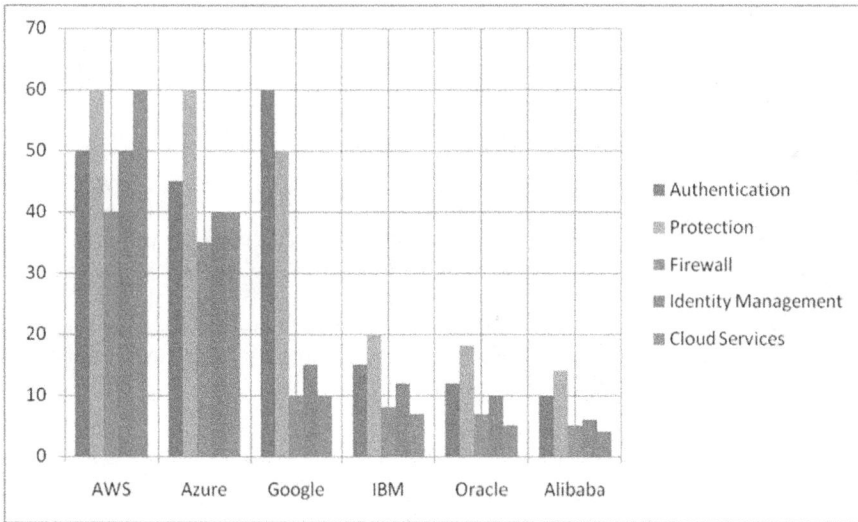

FIGURE 4.3 Security requirements.

of cloud security have been covered in this book chapter. Although cloud computing has several advantages, it conjointly has several drawbacks. As data are stored in large data centers around the world, they can become a target for the hacker to attack or be exploited by cloud computing providers' employees. Furthermore, since the data are stored in several locations, they may be exposed to the conditions with which the owners are unfamiliar. The IT industry is lagging behind, when it comes to cloud computing.

REFERENCES

1. Suresh Kumar, K., A. S. Radha Mani, S. Sundaresan, and T. Ananth Kumar. "Modeling of VANET for future generation transportation system through edge/fog/cloud computing powered by 6G." In: Singh, G., V. Jain, J. M. Chatterjee, and L. Gaur (Eds) *Cloud and IoT-Based Vehicular Ad Hoc Networks*, pp. 105–124. Hoboken, NJ: John Wiley & Sons (2021).
2. Kandukuri, B. R., and A. Rakshit. "Cloud security issues." *In 2009 IEEE International Conference on Services Computing*, pp. 517–520, IEEE (2009).
3. Chitturi, A. K., and P. Swarnalatha. "Exploration of various cloud security challenges and threats." *In Soft Computing for Problem Solving*, pp. 891–899, Springer, Singapore (2020).
4. Popović, K., and Ž. Hocenski. "Cloud computing security issues and challenges." *In The 33rd International Convention MIPRO*, pp. 344–349, IEEE, 2010.
5. Kuyoro, S. O., F. Ibikunle, and O. Awodele. "Cloud computing security issues and challenges." *International Journal of Computer Networks (IJCN)* 3, no. 5 (2011): 247–255.
6. Adimoolam, M., A. John, N. M. Balamurugan, and T. Ananth Kumar. "Green ICT communication, networking and data processing." *In Green Computing in Smart Cities: Simulation and Techniques*, pp. 95–124, Springer, Cham, (2021).
7. Ahmed, M., and M. Ashraf Hossain. "Cloud computing and security issues in the cloud." *International Journal of Network Security & Its Applications* 6, no. 1 (2014): 25.

8. Kumar, K. S., T. Ananth Kumar, A. S. Radhamani, and S. Sundaresan. "Blockchain technology: An insight into architecture, use cases, and its application with industrial IoT and big data." In *Blockchain Technology*, pp. 23–42. Boca Raton, FL: CRC Press (2020).

9. Rao, B. T. "A study on data storage security issues in cloud computing." *Procedia Computer Science* 92 (2016): 128–135.

10. Nazir, M. "Cloud computing: Overview & current research challenges." *IOSR Journal of Computer Engineering* 8, no. 1 (2012): 14–22.

11. Xu, Y., and Paton, D. AWS re:Invent - NEW FEATURES and takeaways (2019). https://medium.com/adobetech/aws-re-invent-2019-new-features-and-takeaways-9b6c7796f435.

12. Li, Z., F. Millar-Bilbao, G. Rojas-Durán, and S. Ladra. "Take one for the team: on the time efficiency of application-level buffer-aided relaying in edge cloud communication." *Journal of Cloud Computing* 10, no. 1 (2021): 1–18.

5 Security Concerns of Virtualization in Cloud Computing

Darshana A. Naik, G. Shruthi,
A. Ashwitha, and Pramod Sunagar
M. S. Ramaiah Institute of Technology

CONTENTS

DOI: 10.1201/9781003219880-5

5.1 INTRODUCTION TO CLOUD COMPUTING

Cloud computing is an on-demand service model focused on the virtualization and distributed computing technologies for IT provision [1]. It's a wide field in the current era. It provides a variety of services and platforms. It provides multi-tenancy, massive scalability, elasticity, and self-sufficiency of resources. Cloud resources are shared so that they can be easily used whenever we want. It offers distinct models of service delivery. They are known as software as a service (SaaS), platform as a service (PaaS), and infrastructure as a service (IaaS). On a subscription basis, SaaS rents apps. The customer should use an approved computer to access the service. In short, SaaS is a model of software delivery in which the application resides on the cloud service provider (CSP) and is accessible through a web browser for its customers (e.g., Google Docs). PaaS provides application developers with an ecosystem for development. So it refers to the distribution over the Internet of operating systems, related instruments, toolkits, and building blocks. A client deploys the application without installing any tool or platform on their local machines on the cloud service provider. The processing power, storage, network, and all other computing resources are outsourced by CSP in the form of virtual machines in IaaS. At this step, a hypervisor is given. Cloud computing provides distinct models of implementation. They are the public cloud, hybrid cloud, and private cloud. Third-party providers host, run, and maintain public clouds. Their protection and regular management are often performed by suppliers. So it is free to all customers. However, private clouds are limited to any enterprise or business of which the company owns networks, infrastructures, and data centers. The hybrid cloud combines sensitive technologies in the private cloud with non-sensitive applications in the public cloud.

The work is structured as follows: Section 5.2 presents the overview of virtualization in cloud, types of virtualization, and different security aspects and security benefits. Section 5.3 outlines the different security challenges in virtualization. Section 5.4 illustrates the common types of virtualized security features. Section 5.5 briefs the cloud-based attacks on the virtualized system. Section 5.6 outlines the best practices to follow for the security in cloud. Finally, Section 5.7 concludes the work with findings.

5.2 VIRTUALIZATION OVERVIEW

Virtualization is a structure used to enhance the use of figuring assets either through the collection, or through isolating the current framework practically. In this cycle, the real or introductory design of the framework is covered up or concealed from the end client. This procedure empowers equipment autonomy, confinement, embodiment, and similarity. This chapter presents an examination of different virtualization advances and methods by and by. This examination basically centered equal virtual machines on a solitary work area that brings to the virtualization climate. The virtual machine (VM) execution is assessed dependent on its information recovery across the stage, security arrangements, and checking measure. Notwithstanding, this audit chapter is an after-effect of the overview of 60 examination articles that were distributed in the main worldwide diaries. The goal of this survey is to investigate the virtualization world from the authors' eyes. Also, this chapter aims to provide a critical understanding of what already exists in virtualization, what needs to be done in virtualization, and the implementation of virtualization environments [2].

5.2.1 FULL VIRTUALIZATION

Full virtualization is a sort of virtualization which is utilized to give a VME that completely reestablishes the equipment hidden it. Any program equipped for executing on the actual equipment can be run on the VM in this kind of climate, and any operating system upheld by the fundamental equipment can be run on every individual VM. Clients can run a few separate operating systems simultaneously. The VM stimulates sufficient hardware in full virtualization, which permits an unmodified operating system that can be run in segregation. In a variety of cases, this is particularly helpful. For example, they are running a test new code in operating system improvement simultaneously as more established variants, each on an alternate VM. VM can be furnished by the hypervisor with all machine assets, along with virtual profiles, virtual gadgets, and the executive virtualized memory. Guest OS is completely disengaged by the virtualization layer from the underlying hardware [3]. By utilizing a blend of twofold interpretation and direct execution, complete virtualization is accomplished. The actual central processor executes sensitive directions at local speed with complete virtualization hypervisors. Operating system directions can be deciphered on the ground and put away throughout time, and client-level guidelines run unlimited at local speed. Full virtualization gives the best VM disconnection and assurance and improves on movement and conveys ability, as virtualized or local equipment may run a similar operating system case.

5.2.2 PARAVIRTUALIZATION

Paravirtualization can be changed by OS, which involves kernel. The preprocessing can be done between the programs where the OS kernel acts as a bridge at a level of difficulty. Paravirtualization replaces hypercalls that interact directly with the hypervisor virtualization layer for non-virtualizable instructions. A call to the computer depends on the same documentation as a hypercall. Solicitations are settled on framework decisions to get to application assets from the operating system, and the interface can be given between the application or technique and the operating system. Hypercalls, except that the hypervisor is used, function the same way. The hypervisor also provides other kernel operations with hypercall interfaces, which include memory management and interrupt handling. In paravirtualization, the VM emulator is initiated after the host OS boots. VMLAUNCH (Intel) or VMRUN (AMD) is used by the emulator instructions to start VM execution, depending on the platform. In existence, there are two copies of OS at this point. While the guest runs in an active state, in suspension mode, the host plays a very important role. While paravirtualization in production environments can bring support problems since it needs deep OS kernel modifications, compared to full virtualization, it is relatively simple. An instance of paravirtualization is the open-source Xen Project.

5.2.3 TRADITIONAL SECURITY APPROACHES TO VIRTUALIZATION

There have been security problems with virtualization in recent years, and we have studied and documented security problems contained in distinct elements of cloud virtualization, for instance, hypervisor, virtual systems, and guest disk images. The best example to exploit is a hypervisor-based attack, an attacker who exploits security flaws to make a single hardware processor be shared by multiple operating systems. A hypervisor that is compromised will cause the hacker to target a virtual host on each virtual machine. The risk is increased by a greater number of APIs and program stacks, resulting in a lower level of code protection assurance. In virtualized environments, we emphasize the following assaults [4].

5.2.3.1 VM Escape

In order to promote a strong separation between the host and the VMs, virtual machines are designed. A malicious program can be injected by the attackers, which is run by VM. VM removes the isolated boundaries when this program is run and begins to connect with the help of the operating system, directly avoiding the VMM layer. On the host computer, further attacks can be gained and conducted by an attacker, which acts as an exploit to the environment.

5.2.3.2 Hyperjacking

A good example of an attack is hyperjacking, inside a virtual machine (VM) host; the hypervisor is controlled by the hacker, which generates the free thesaurus. The attack's goal is to focus on the working framework underneath virtual machines to permit the program of the assailant to run, and its essence will be straightforward to the VM applications above it.

5.2.3.3   VM Sprawl

VM spreads occur on account of tremendous measure of virtual machine live in the framework without legitimate observing or control. Some of the examples of device assets are memory, disks, network interfaces, etc., and are maintained during this time, and these assets cannot be allocated and are essentially lost to other VMs.

5.2.3.4   Hypervisor Vulnerabilities

The hypervisor is created between guest VMs. Multiple guests can be run, and the applications can be run on a single host. Compromising a VMM also allows the physical device and the hosted applications to gain power. For example, Blue Pill and hyperjacking inject rootkits based on virtual machines (VMs) that can be set up or replace the prevailing rogue hypervisor to gain 100% control over the situation. Since the hypervisor is implemented below the host OS, standard security measures make it difficult to detect these types of attacks.

5.2.3.5   Single Point of Failure

The infrastructure that controls VM access to primary resources is the foundation of today's virtualized environments and a vital machine feature. As a result of the failing hypervisor, the entire machine collapses due to overworked hardware or software culpabilities.

5.2.3.6   Inside-VM Attack

The VM can be contaminated at runtime, which contains malware or OS rootkits. VM is controlled by the attack and seriously compromises its security state.

5.2.3.7   Outside-VM Attack

The host OS attacks and assembles VMs and are referred to as "elevation-VM attacks." Elevation-VM attacks are difficult for clients to counter, which is by means of shared memories, association with network, and other shared resources; a malicious VM may potentially access other VMs. For instance, if a malevolent VM decides where the granted memory of another VM lies, then it may do the reading or writing operation to that location and interfere with the operation of the other VM.

5.2.3.8   Cross-VM Side Channel Attack

A few VMs are regularly introduced on the essential of a cloud-based worker to upgrade asset use, and this co-residence situation is a budding test to cross-VM side channel assault. The fundamental initiative is a fraudulent VM that spikes the detachment among VMs, and afterward accesses the common equipment and stores positions to remove touchy data from the objective VM.

5.2.3.9   Outdated SW Packages in VMs

In a virtualized world, obsolete digital computer program packages can pose severe security risks. For instance, a system rollback operation can reveal a software bug that has already been patched. Such operations can have significant security implications.

5.2.3.10 VM Footprint

VM footprint is the capability used to collect target VM information, such as installed OS, established packages and running assistance, etc. Clients like to make new virtual machines for different assignments due to the practicality and the effortlessness of development, fanning out new virtual machines dependent on the old ones, preview machines, or uniform-moving machines to an untimely state.

5.2.3.11 VM Image Sprawl

In order to construct new virtual machine image files, live virtual machine images can simply be duplicated, which can leave unseen after-effects in VM image sprawl problem wherein a large number of virtual machines created by cloud consumers can be left hidden. VM picture has implications for the enormous administration problems of security patch VMs.

5.2.4 SECURITY BENEFITS DUE TO VIRTUALIZATION

Virtualization is the kind of software that should be applied with intention, a particular purpose in mind. These priorities can include, depending on the organization, everything from reducing capital expenditure in hardware to increasing the efficacy of existing IT systems. Security can be the fuel behind your virtualization dreams if network interference and data breaches are issues that keep you up at night. Network virtualization, ever so versatile, will enhance security in many ways. Some explanations are here.

5.2.4.1 Centralizing Confidential Data

Centralized control is where more companies are pursuing to counteract one of the main organizational challenges of today: the careless worker. Statistics indicate that missing laptops cost businesses about $50,000 on average, which accounts for the cost of data breaches, intellectual property damages, and replacements. Virtualization provides solid protection against fraud, virus infections, and other misfortunes that can come the way of the employee by allowing administrators to centrally manage data.

5.2.4.2 Providing Secure Access

The introduction of mobile phones and tablets has mobilized the workforce by allowing employees to be involved at home, the nearby cafe, and other locations other than the office. As a result, enterprises of all different magnitudes are scrambling to find solutions to ensure that employees from around the globe have the right access to use the apps and the results. Virtualization allows workers instant access to company services from anywhere through a centrally controlled and secured virtual network back at the workplace.

5.2.4.3 Setting up a Sandbox

Employers are correct to be apprehensive about allowing employees to access the web from company networks, with all the malicious apps, hackers, and scam artists running amok. The definition of the sandbox is all about isolating applications that could pose a network hazard, and the ideal environment is provided by virtualization. For example, to build one wide sandbox that is solely designed for web surfing, you can use a hypervisor. This way, inside the virtual world, any attacks are isolated without damaging the host server and the other applications that it may run.

5.2.4.4 How to Tackle Virtualization Security Challenges

Virtualization does a lot to assure security, but all new security threats can be added simply by using it. Being isolated ensures that security threats will avoid and wreak havoc across the targeted host from existing security mechanisms. Anything from DOS attacks to data breaches can compromise the critical aspects of output within, and probably outside, the virtual world if precautions are not taken. The following guidelines are intended to mitigate network virtualization-related gambling.

5.2.4.5 Know the Risks

The fact that virtualization comes with risks has already been developed, but if you intend to implement it, you need to understand the risks it poses to your company. While particular risks can vary depending on a variety of factors when you:

- keep login credentials on the host computer or other confidential information,
- enable online access to virtual machines,
- allow other networked computers and devices to access virtual machines, and
- move virtual computers between real host computers.

Conducting a risk assessment will help you recognize the company's most important security issues and find the solutions required to resolve them.

5.2.4.6 Harden the Host Machine

Handfuls or many virtual machines might be controlled by a solitary worker. It's a staggering accomplishment, and there's a ton to get some information about the equipment also. The host machine, particularly the working framework, should be firmly rushed down for this reason. Ensure that it is fitted with everyday refreshes, antivirus applications, firewalls, and all the security applications that are suggested. It can be a guarantee that VMs have a strong layer of safety directly out of the door by reinforcing the host stage.

5.2.4.7 Police Network Access

Designing the right network access policies is a big part of resolving the issues of virtualization security. Not everyone who has access to a virtual machine needs permission to start, install, and update such machines. Similarly, not every VM actually needs Internet connectivity. Strict access policies must be implemented, from accessing individual programs to security settings, to ensure network privacy. Unplug what you're not using like physical machines; their use can be exhausted over time by virtual machines. If you can help, avoid deploying VMs that do not give business operations any specific reason or benefit. Once their welcome is worn out, they should be removed from the network and all linked devices, virtual or otherwise, immediately. The longer these non-functional devices stay in circulation, you run the risk of disrupting the production environment.

5.2.4.8 Be Realistic

Those desktops or servers that you crank out of the hypervisor might be virtual, but physical machines have much of the same needs. That means it's always important to install updates and patches. In circumstances where an enterprise only fires up VMs on an as-needed basis, automated delivery is feasible, but problematic. Whatever the case, IT should be completely aware of all of the network's virtual networks to ensure that security staff receives the necessary attention. To allow horror stories about security to serve as adoption barriers, virtualization offers far too many advantages. Yeah, it takes some work, but while keeping the risks to a minimum, it is possible to experience the best this technology has to offer. Know the situation, know the risks, and most of all, take to heart the best practices.

5.3 SECURITY CHALLENGES IN VIRTUALIZATION

Along with advantages, there are many disadvantages of virtualization. The implementation of virtualization will add an extra cost to the customers, and the customers or companies must adapt themselves to learn new infrastructures and implement them. With virtualization, organizations achieve better management of services offered. But it also brings critical security issues for the organizations [5]. A few of the key security concerns are itemized below:

- **Active and inactive guest images:** The guest images can be available in active and inactive modes. Both are vulnerable to security issues. Active VMs are vulnerable to any regular kind of attacks that are seen on any connected devices. Inactive VMs may not be having the updated security patches and hence are vulnerable to attacks.
- **Attack on the host OS:** The guest VMs when run on the host OS are vulnerable to attacks if the host gets compromised. The chances of the host getting compromised are rare, but new vulnerabilities may have an adverse effect, and the system may be attacked. An intruder using guest VM can exploit vulnerabilities to increase their permissions to access the server, and once the attacker has hold of the host, he will manipulate the entire host VMs at the level of the host. A guest-to-host strike is the term for this situation. Another type of attack is a denial of service attack, which involves sending too many bad requests to the web server, causing it to stop serving legitimate requests.
- **Attack on the hypervisor:** The attack on the hypervisor is carried out through VM escape. Here the attacker is attempting to crack the security layer of the OS and initiate communicating unswervingly with the hypervisor. The hypervisors are updated numerous times at runtime to prevent downtime in the cloud architecture. Backdoor or rootkit can be introduced into the hypervisor during this modification and can be used later to monitor the entire system.

5.4 TYPES OF SECURITY IN VIRTUALIZATION

There are several virtualized security characteristics, which include network security, application security, and cloud-based security. Virtualized versions of conventional security technology are radically modified (i.e., next-generation firewalls). Others are revolutionary innovations integrated into the virtualized networks.

Common types of virtualized security features [6] include the following:
- Segmentation
- Micro-segmentation
- Isolation.

5.4.1 SEGMENTATION

Segmentation or the making available of specific services only to specific applications and users generally takes the form of managing traffic between various segments or levels of the network.

5.4.2 MICRO-SEGMENTATION

Micro-segmentation allows programmers to allocate fine-grained security policies down to the level of workload. Two core concepts are built around it: granularity and dynamic adaptation. Micro-segmentation is radically different from traditional network segmentation by the implementation of these concepts. Granular segmentation tracking and compliance points that are readily accessible and cheap are the tricks to making segmentation fine. In the network, in the virtualization platform hypervisor layer, on hosts, on cloud services, remember that a little extra granularity can go a long way. Just by fencing off a few high-value applications or separating development and test systems from production environments, you can dramatically improve protection. Dynamic segmentation in today's dynamic environments, if administrators had to build and maintain security rules at each compliance stage manually, granular segmentation would be impractical.

More compliance points will lead to less flexibility, more chances for error, and incredibly long working days for security and network managers. By integrating abstraction, intellect, and automation, "dynamic segmentation" makes a granular approach to segmentation possible. Abstraction is the ability to convey security policies in terms of application concepts rather than network structures (such as site, application, and database levels) (such as IP addresses, subnets, and VLANs). Because of hairpinning architectures, conventional appliance-based approaches to securing internal traffic generate uncertainty, decreasing developers' resilience in implementing distributed applications and contributing to safety compromises. Micro-segmentation offers an agile, least-privilege security model central to modern data centers described by software instead of implementing several physical firewalls [7]. Intelligence can identify when applications or infrastructure changes are made and then reconfigure policies to respond to the changes. Automation is the capacity, without human involvement, to rapidly deploy new and updated security policies to monitoring and compliance points. Micro-segmentation can also improve addictiveness in other ways, depending on the implementation [8].

5.4.2.1 Benefits of Micro-Segmentation

Organizations that implement micro-segmentation recognize tangible benefits.

More specifically:

- **Reduced attack surface**: Without slowing growth or creativity, micro-segmentation offers insight into the whole network climate. Early in the design and development phase, program developers should integrate the security policy theory to ensure that application implementations or revisions do not create new attack vectors, which is especially critical in DevOps' fast-paced world.
- **Improved breach containment**: Micro-segmentation uses security teams to track network traffic against predefined policies, which reduces the time taken to react to and fix security gaps.
- **More substantial regulatory compliance**: Micro-segmentation allows officers to create policies that isolate the systems subjected to regulations from the rest of the infrastructure, reducing the likelihood of non-compliance by granular monitoring of interactions with regulated systems.
- **Streamlined policy management**: Instead of executing these functions in distinct parts of the network, switching to a micro-segmentation architecture simplifies firewall policy management. This technique decreases the threat surface and strengthens the enterprise's defense position.

5.4.3 ISOLATION

Pr. Stuart Madnick and JJ incorporated the use of virtualization technology for defense when working on Project MAC in 1973, Donovan (MIT) ([MD73]). The hypervisor is also used as a reference monitor in order, among other things, to isolate the workload within a VM, while OS provides only poor isolation between processes. This capacity of isolation allows the load balancing of the hosting machines [9].

There are four distinct methods used in VM isolation.

1. Level 1 hypervisor's lifetime kernel code integrity.
2. Any weather violations in the above three parameters must be reported to the guest VMs running in Level 2. External attacks are not feasible with a level 0.0 hypervisor, according to Stable Turtles.

It is not very practical to provide each VM with individual I/O devices. As virtual I/O devices are assigned, it won't be easy to ensure separation between virtual machines without a hypervisor [10].

5.5 ATTACKS ON THE VIRTUALIZED SYSTEM

5.5.1 HYPERVISOR-BASED ATTACKS

The most common form of virtualization used in the cloud is hypervisor-based virtualization because it enables intrusion detection systems. The hypervisor is directly mounted on the hardware in this process. The manager VM manages all the other VMs present on the system made up of a single virtual machine. The hypervisor is the single failure point for this virtualization technique. Gaining hypervisor control

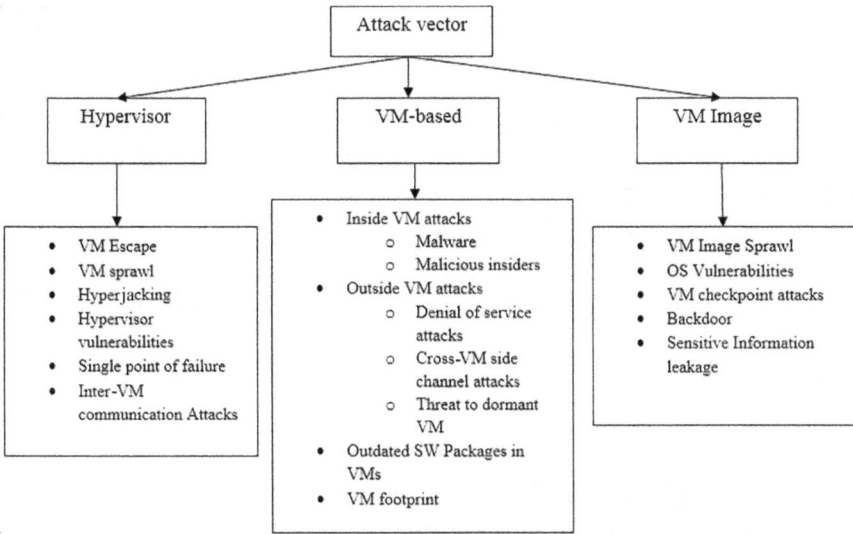

FIGURE 5.1 Taxonomy of cloud-based attacks.

enables all virtual machines to be managed. A hypervisor-based attack is one kind of attack in which an attacker takes advantage of programmer flaws to cause several operating systems to divide a typical hardware processor among them. A vulnerable hypervisor could encourage a hacker to target each virtual machine's virtual host [11] (Figure 5.1).

5.5.1.1 Types of Hypervisors

The two main types of hypervisors (or VMM) are as follows:

- Type I hypervisors (or native or bare-metal hypervisors) are the actual host's operating software or hardware. They function as a VM operating system and monitor access to the existing hardware directly.
- Type II hypervisors (or host hypervisors) are software that runs above a conventional operating system, essentially an operating procedure on the natural host's OS. At the same stage, other processes will coexist. Therefore, guest OS and VM are above the layer (a third one) than the hypervisor [12] (Figure 5.2).

Types of hypervisor-based attacks:

- **VM escape:** Virtual machines are premeditated and developed to maintain a robust separation between the host machines and the VMs [13].
- **Hyperjacking:** Hyperjacking is an assault in which an attacker gains negative control of the hypervisor, which produces the virtual world within a virtual machine (VM) host. The attacker intends to attack the operating system below that of the virtual machines for the attacker's software to run unnoticed by the applications running on the VMs [14].

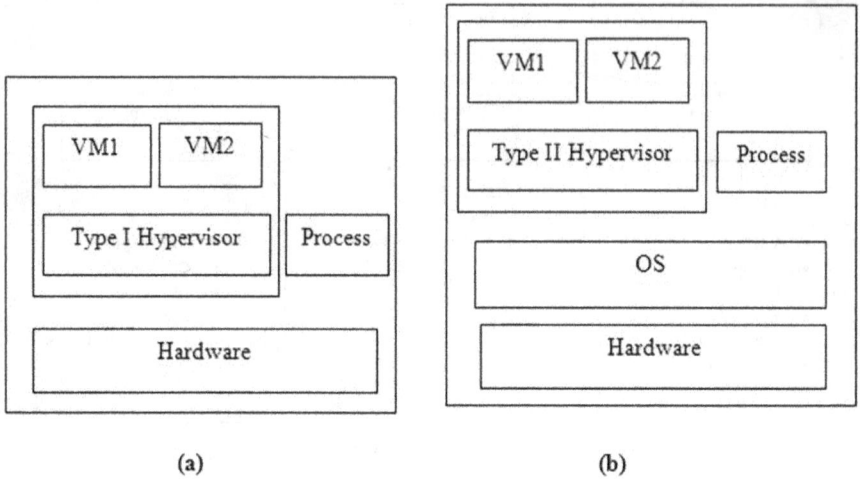

FIGURE 5.2   Type I and type II hypervisors [8].

- **VM sprawl:** If there are so many virtual machines in a system and there isn't enough control or management, VM sprawl occurs. Since computer resources (such as memory, disks, and network channels) are retained throughout this period, they cannot be allocated and are effectively lost to other virtual machines.
- **Hypervisor vulnerabilities:** A virtual machine manager (VMM) is a program that runs multiple guest virtual machines and multiple applications on a single device while maintaining separation between them [15].
- **Single point of failure:** The current virtualized environment controls virtual machines' connection to physical resources and is required for the device's overall functionality. As a result, hypervisor failure caused by overused hardware or software faults causes the entire system to crash.

5.5.2   VM-Based Attacks

- **Inside-VM attack:** The virtual machine may become infected with malware and infected by OS rootkits while running. This type of attack will take complete control of the virtual machine and seriously jeopardize its stability.
- **Outside-VM attack:** Outside-VM attacks are attacks on the host operating system and co-located virtual machines. Consumers have a hard time defeating external VM attacks. A malicious VM can access other VMs through shared memory, network connections, and other shared resources. For example, if a malicious VM decides another VM's allocated memory location, it can read or write to that location, interfering with the other VM's operation [16].

- **Cross-VM side channel attack:** Several VMs are typically deployed on the same physical server to maximize resource utilization in the cloud. Cross-VM side channel assault can be hampered by this co-resident placement [17]. The underlying principle is that a fraudulent virtual machine (VM) breaks through virtual machine isolation and then uses pooled hardware and cache to steal confidential data from a specific VM [18].
- **Outdated SW packages in VMs:** Some rare and vital software packages on virtual machines can pose serious security risks in today's virtualized environment [11]. Because of the minimal effort and comfort of improvement, clients decide to fabricate new virtual machines for different assignments dependent on old virtual machines, depiction machines, or even rollback machines to an initial state [19]. A device rollback, for example, may expose a previously patched programmer error. Such operations may have serious security consequences.
- **VM footprint:** Virtual machine footprint is a method for collecting data on the target virtual machine, such as the installed OS, installed packages, and running services. It refers to one of the pre-attack stages or the tasks performed before the actual attacks [20].

5.5.3 VM IMAGE ATTACKS

- **VM image sprawl:** Virtual machine images taken in live are reproduced to create fresh virtual machine image files, leaving unseen after-effects in the VM image sprawl issue, where a large number of VMs created by cloud consumers can go unnoticed. The VM image spreads implications in significant administration problems of VMs containing security fixes.
- **VM checkpoint attacks:** Virtual machine images are mostly online, and they'll have to start repairing them. That will help to cover the computing costs of the cloud service provider. An intruder may try to get to the VM checkpoint on the disk containing the physical memory, VM content, and sensitive VM state data, exposing sensitive VM state data. The cloud computing virtualization domain has a far larger attack area than the traditional computing domain. Various types of cloud storage service models have varying stages of security services. Users of the cloud get the least integrated security with the infrastructure-as-a-service provider and the most with the software-as-a-service provider.

5.6 BEST PRACTICES

5.6.1 NETWORK SECURITY

Providing protection to network is some action that a body takes to prevent the personal information of the network, with various consumers, or their system from being maliciously used or inadvertently harmed. The aim of network protection is for all legitimate users to keep the network running and secure blocking unauthorized network access. As there are so many places for a network to be vulnerable, network protection encompasses a broad range of operations. They include the following:

- **Organizing powerful machines:** Using tools to prevent malicious programs from accessing the network or running within it. Blocking users from sending or receiving e-mails that appear suspicious. Blocking unauthorized network access. It also prevents users of the network from accessing websites considered to be harmful.
- **Passive system deployment:** For example, utilizing devices and applications that detect unwanted network intrusions or unusual behavior by approved users.
- **Usage of preventive technology:** Devices that help find possible gaps in defense.
- **Ensure that users follow best practices:** Even though the software and hardware are designed to be safe, users' actions will create security holes. Network security professionals are responsible for training the organization's members on how to remain safe from external threats.

5.6.2 ESSENTIAL NETWORK SECURITY BEST PRACTICES

Taking a layered approach to the protection of your company is important [21]. To develop a systematic network security management plan, adopt these ten cybersecurity best practices (Figure 5.3).

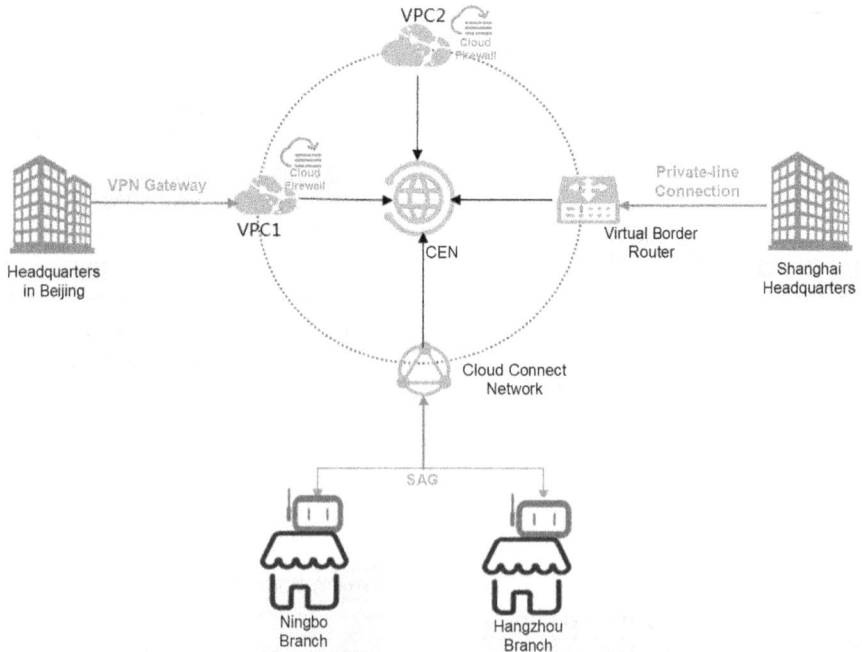

FIGURE 5.3 Best practices for network security [22].

1. Implement a systematic approach to the governance of IS
2. Data loss stop
3. Insider hazard identification
4. Information backup
5. Look out for societal business
6. Train the apps
7. Delineate obvious utilized policies for innovative workers
8. Computer and systems upgrade
9. Playbook build an occasion retort
10. Compliance to maintain.

Disaster recovery is the mechanism by which normal activities are resumed by regaining access to data, hardware and software, networking equipment, electricity, and communication after a disaster. However, operations may also apply to practical issues such as seeking alternative workplaces, repairing communications, or sourcing everything from desks and computers to employee transportation if the facilities are damaged or demolished. The response to disaster recovery should adopt a recorded practice or usual of processes precisely intended to arrange the association aimed at retrieval in the unswerving conceivable situation, a disaster salvage strategy.

5.6.2.1 Disaster Recovery Stages
In the following sequential phases, disaster recovery happens:
1. **Establishment stage**: The disaster achieves are evaluated and reported here.
2. **Completion stage**: In this phase, the actual procedures are performed to restore each of the organizations impacted by the disaster. In the recovery system, company processes are restored.
3. **Reconstitution process**: The original system is restored in this phase, and procedures for the execution phase are stopped.

   3.1 **Activation process with or without warning**: A disturbance or emergency can occur. Illustrations such as a storm impacting some certain topographical area, or a disease eruption predicted at a convinced day. There can, however, remain not at all notice of the burst of an aquatic channel or of a criminal performance by humans. The secret to mitigating the consequences of the inward alternative is the quick and effective identification of a calamity incident and having an adequate message strategy; in few situations, it will provide ample period toward consent organization team to gracefully execute achievements, thereby diminishing the bearing of the adversity. The adversity regaining commission is responsible aimed at the motivation process being initiated. The topographical, dogmatic, societal, and conservational activities could cause intimidations to the commercial processes of the organization be well aware. To forestall untrue fears or exaggerates to swindles, it ought to take reliable intelligence foundations in the numerous agencies.

   The activation method requires:
   - Announcement of protocols
   - Assessment of impairment

- Preparation for operation for business continuity.

3.1.1 **Damage assessment**: In order to decide whether the eventuality strategy would be performed after a power failure, it is critical to analyze the extent and gradation of impairment to the infrastructure. This assessment of impairment would be carried out as rapidly as finances allow, with the highest priority assumed to the security of workers. The disaster response squad is also the primary squad summary of the team.

Source of disturbance or emergency:

1. Potential for further intervention or harm
2. Area affected by the emergency
3. State of educational environment
4. Stock and functioning status of the most main characters
5. Form of failure of equipment
6. Items which should be substituted.

5.6.3 PROTECTION OF A VIRTUAL MACHINE

The virtual machine service provider must implement software applications that prohibit the use of financial capacity by virtualization as sanctioned. Additionally, a frivolous procedure necessity track happening a virtual machine that folds kindling after the VMs display them in everyday life in order to repair any VM tampering (Figure 5.4).

The shown practices for virtual machine protection ensure the safety of exploitation:

1. General defense of virtual machines can be made use.
2. Using virtual machines deployment models with baseline level of safety.
3. Minimize virtual machine console use.

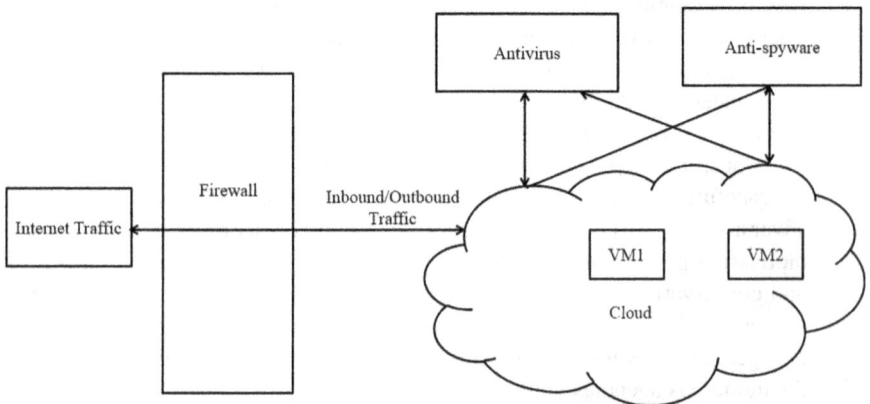

FIGURE 5.4 VM securities by firewall, antivirus, and anti-spyware.

4. Preventing the takeover of resources by virtual machines with host resource management features.
5. Disable unnecessary features within virtual machines.

5.6.4 THE SYSTEM OF MANAGEMENT

Management of virtualization is software that integrates with virtual worlds and the underlying physical hardware in order to simplify the management of resources, improve data processing, and streamline operations. Each virtualization management system is special, but most feature an uncomplicated user interface that streamlines the process of creating a VM, monitors virtual environments, allocates resources, compiles reports, and enforces rules automatically. Some solutions also combine through brands of hardware and software, enabling users to install the best management system for them.

5.6.5 SECURITY HYPERVISOR

During its life cycle, a hypervisor compliance remains the practice of guaranteeing that the hypervisor, the software that allows virtualization, stands protected. This includes during production and during execution. Curtailing local device users, boundary spell exteriors in addition retaining completely coordination informed are common hypervisor products or services. A hypervisor is a design pattern that separates the browser. Hypervisor defense strategies and best practices are listed below.

- There are various methods to help secure hypervisors, with operations, for example by means of external tools for displaying in addition system safety, diminishing occurrence exteriors, scenery admittance rights, and informing the hypervisor in addition possession the corporeal attendant obtainable of unsanctioned people's control.
- Finished means of tackles aimed at monitoring and network security, superintendents resolve remains capable toward footpath their virtual environments and detect any apprehensive behavior early on. Safety tools from vendors such as Planetary Storms, Inc., and VMware help track threats in addition to identifying them. Extra software for network security tools, such as firewalls, can be used by administrators.
- Diminishing possible occurrence exteriors resolve type is harder toward admitting a computer-generated computer through declining possible ideas. Numerous functioning arrangements resolve preservative characteristics that an entity that will increase the attack surface of a VM might not be essential. To reduce hazard surfaces, superintendents can disable unnecessary services and only allow needed services.
- The coordination superintendent must likewise employ usual limitations on who is able to remotely from the support contact the hypervisor. There resolve similarly boundary who container monitor the surroundings of the hypervisor in addition avoid unsanctioned manipulators since changing or

retrieving evidence. The greatest hypervisor boards sustenance numerous contacts.
- A further standard procedure is to restrict physical server access,since a reasonably easy way of accessing the hypervisors is connected to the storage computer.

5.6.6 REMOTE ACCESS

Remote access is the action of involving IT facility, software, or information to a position further than head office or a locality contiguous to the information center. This assembly lets manipulators contact the system or device distantly through a network assembly or a telephone connection.

Safe remote access advantages also provide protected distant admission advantages, which are as follows:
- Improved employment quality in addition to elasticity
- Improved end-to-end coverage with BYOD and lesser cost
- Enhanced market steadiness.

5.6.6.1 Good Practices

Many companies are encouraging workers to operate remotely with the help of cloud technology. Being able to work from anywhere, after all, increases versatility and efficiency. As they aim to ensure smooth access to resources and software, other significant programs, including essential security-related measures, are being placed on the back burner. Cybercriminals notice that businesses are more vulnerable than ever.

Checklist 1: Make remote configuration stable regardless of positioning up the communications.

- Using controlled devices wherever possible.
- BitLocker for Windows and FileVault for MacOS to allow encryption.
- Install antivirus and firewall security.
- Ensure that the supplier currently supports both operating systems and other applications.
- Implement a high-quality strategy on passwords, disable automatic login, and allow automatic locking.
- Allow "find my device" with capabilities for distant security device/clean.

Checklist 2: Minimize the peril of extended surface assault.

It will help make your environment safer by taking the measures in the previous checklist. Apply the shown strategies to strengthen risk management further. Opt basic best practices for housekeeping. Particularly:

- Classify all accounts that are musty and then uninstall them.
- Verify all acquiescence and delete unused and unnecessary rights.
- Trim the records with rights.
- Revising the model with allocation.

- Shut down the never utilized network repairs or remove them.
- Distill your strategy for the party.
- For access control through your infrastructure, use Active Directory and Azure AD classes. In order to ensure that no one has undue permissions, check the groups and group membership periodically.
- Ensure that NTFS permissions and permissions comply with the least-privilege principle for shared tools.

5.7 CONCLUSIONS

Virtualization involves the use of hypervisor or virtual machines to surround an operating system and provides the alike set of functionalities that are expected from an actual hardware device. Virtualization links VMs to physical infrastructure, but raises protection issues when users bow out physical control of their computers and files. Virtualized systems create major gaps in the security that needs to be carefully considered when implementing a sturdy and safe and sound virtualized environment. The multi-tenant essence of virtualized structures must be handled carefully to maintain good segregation within the tenants' activities. The environment of the virtualized cloud can be exploited by a number of hypervisor hacks, virtual machines, and VM images. A detailed comparative analysis is presented on the different best practices suggested by research experts to tackle vulnerabilities to virtualization in the cloud computing framework. The future scope of the virtualization technology involves implementing a hybrid IT system to provide an end-to-end solution. Virtualization technology should strive to provide the security to the user's data. The availability of the application all the time on cloud and also its maximum utilization is another important aspect that needs to be addressed for efficient utilization of the resources.

REFERENCES

1. Tank, D., A. Aggarwal, and N. Chaubey. "Virtualization vulnerabilities, security issues, and solutions: a critical study and comparison." *International Journal of Information Technology* (2019): 1–16.
2. Fayyad-Kazan, H., L. Perneel, and M. Timmerman. "Full and para-virtualization with Xen: a performance comparison." *Journal of Emerging Trends in Computing and Information Sciences* 4(9) (2013): 719–727.
3. Fathima, K. M. M. and N. Santhiyakumari. "A survey on evolution of cloud technology and virtualization." *In 2021 Third International Conference on Intelligent Communication Technologies and Virtual Mobile Networks (ICICV), Tirunelveli, India,* pp. 428–433, IEEE (2021).
4. Sugumar, T. N. and N. Rajam Ramasamy. "Literature survey and review: virtualization technologies." December 2016.
5. Kumar, V. and R. S. Rathore. "Security issues with virtualization in cloud computing." *In 2018 International Conference on Advances in Computing, Communication Control and Networking (ICACCCN), Greater Noida, India,* pp. 487–491, IEEE (2018).
6. https://www.vmware.com/topics/glossary/content/virtualized-security/.
7. https://www.vmware.com/topics/glossary/content/micro-segmentation/.
8. https://cdn2.hubspot.net/hubfs/407749/Downloads/Illumio_eBook_The_Defnitive_Guide_to_Micro_Segmentation_2017_08.pdf.

9. Clemente, P. and J. Rouzaud-Cornabas. "Security and virtualization: a survey" (2011).
10. Krishna, S. R. and B. P. Rani. "Virtualization security issues and mitigations in cloud computing." *In Proceedings of the First International Conference on Computational Intelligence and Informatics*, pp. 117–128, Springer, Singapore (2017).
11. Suresh Kumar, K., A. S. Radha Mani, S. Sundaresan, and T. Ananth Kumar. "Modeling of VANET for future generation transportation system through edge/fog/cloud computing powered by 6G." In: Singh, G., V. Jain, J. M. Chatterjee, L. Gaur (Eds.) *Cloud and IoT-Based Vehicular Ad Hoc Networks*. Hoboken, NJ: John Wiley & Sons (2021): 105–124.
12. Adla, V. "Comparing performance of HyperV and VMware considering network isolation in virtual machines." PhD dissertation, Dublin, National College of Ireland (2013).
13. Jansen, W. A. "Cloud hooks: Security and privacy issues in cloud computing." *In 2011 44th Hawaii International Conference on System Sciences*, pp. 1–10, Hawaii, IEEE (2011).
14. Hyde, D. "A survey on the security of virtual machines." Department of Computer Science, Washington University in St. Louis, Technical Report (2009).
15. Kazim, M. and S. Y. Zhu. "Virtualization security in cloud computing." *In Guide to Security Assurance for Cloud Computing*, pp. 51–63, Springer, Cham (2015).
16. Xiong, H., Q. Zheng, X. Zhang, and D. Yao. "Cloudsafe: Securing data processing within vulnerable virtualization environments in the cloud." *In 2013 IEEE Conference on Communications and Network Security (CNS)*, National Harbor, MD, pp. 172–180, IEEE (2013).
17. Schwarzkopf, R., M. Schmidt, C. Strack, and B. Freisleben. "Checking running and dormant virtual machines for the necessity of security updates in cloud environments." *In 2011 IEEE Third International Conference on Cloud Computing Technology and Science*, Athens, Greece, pp. 239–246, IEEE, 2011.
18. Schwarzkopf, R., M. Schmidt, C. Strack, S. Martin, and B. Freisleben. "Increasing virtual machine security in cloud environments." *Journal of Cloud Computing: Advances, Systems and Applications* 1(1) (2012): 1–12.
19. Sosinsky, B. *Cloud Computing Bible*, vol. 762. Hoboken, NJ: John Wiley & Sons (2010).
20. https://www.observeit.com/blog/10-best-practices-cyber-security-2017/
21. https://www.alibabacloud.com/blog/best-practices-for-building-secure-global-networks-internal-and-external_596262.
22. Kazim, M., R. Masood, M. A. Shibli, and A. G. Abbasi. "Security aspects of virtualization in cloud computing." *In IFIP International Conference on Computer Information Systems and Industrial Management*, pp. 229–240, Springer, Berlin, Heidelberg (2013).

6 Operational Security Agitations in Cloud Computing

N. Nithiyanandam and S.P. Priyadharshini
Bharath Institute of Higher Education and Research

G. Fathima
Adhiyamaan College of Engineering

S. Jeyapriyanga
Bharath Institute of Higher Education and Research

CONTENTS

6.1   CLOUD COMPUTING—AN OVERVIEW

The development of computing technology and the increase in data dependency require surplus amount of data to be stored and handled in the server for serving the users when needed. It is really a challenging factor to install a physical infrastructure of data server over huge acres of land, and the maintenance of those physical infrastructures involves variety of challenges such as proper maintenance, resource availability, and above all the framing of security policies and preserving the data

DOI: 10.1201/9781003219880-6

from unauthorized access. The data centers are tied up with handling huge quantity of carbon prints with physical infrastructure and extract more power, space of occupancy, cooling factors, etc. Based on Gartner's analysis report [8], most IT organizations spend about 70% of their budget for installation and maintenance of physical infrastructure. To overcome the aforementioned concerns, it is advisable to migrate to virtualization [2] of infrastructure such that huge quantity of data can be stored in a limited area with high level of security features. This special feature attracts IT professionals and organizations to move their physical infrastructure to virtual background. The introduction of virtualization technology [1] has reduced the hardware utility from 100% to 20% with improved level of security and quantity of data storage. Virtualization has initiated its evolution in the earlier 1990s in the form of virtual machines (VMs) and had migrated to virtual cloud servers, and at present, as per Spiceworks' research [10] reports, around 96% of IT organizations had drifted their business data to virtual platform. Virtualization possesses various salient features, and the need for virtualization is depicted in Figure 6.1.

From Figure 6.1, the following are the highlighted needs for virtualization.

- **Limited resources**: The conventional physical infrastructure is sufficient to satisfy the basic computation needs of a user, whereas considering the quantity of computation data, as the data increase the performance of the computing device will be reduced. The virtual machines have sufficient resources with VM managers that can yield high level of performance with reduced infrastructures. The IT organizations have no need of procuring physical devices and software packages in the virtual platform.
- **Reduced area of occupancy**: Major IT tycoons such as Google, Amazon, Microsoft, and IBM handle huge quantity of data, and in physical infrastructure, it involves huge acres of area, physical computing devices, and programming packages to store the data. Hence, the aforementioned tycoons migrated to virtual platform with the aim of reducing the area of occupancy. This technology overcomes the pitfalls of paying for huge area consumption for additional resource capability.
- **Enhanced performance**: The increase in volume of data will decrease the performance of the computing device such that the processor consumes huge time for data arbitration. The virtual managers in the virtual machines

FIGURE 6.1 Need for virtualization.

procure the data and classify them based on multiple factors such as data usage, type of data, and frequency of access by the users such that the data can be moved from primary, secondary, and tertiary memory locations. This memory classifications ease the VM managers to access the data at a rapid rate to increase the performance.

- **Eco-friendly initiatives**: The massive data centers consume huge power and require surplus cooling mechanism for thermal drifting process. The virtualization reduces the area of occupancy, and hence power consumption is reduced to a greater extent, and a better quantity of energy consumption requires minimized cooling options by means of server consolidation process.
- **Reduced administrative costs**: The increase in demand for the huge capacity of the data centers leads to increased installation and maintenance costs. The process of virtualization reduces the necessity of huge data centers, and hence the number of servers will be reduced as a result of server virtualization process. This process reduces the installation and maintenance costs of data servers.

6.2 CLOUD COMPUTING SERVICES AND APPLICATIONS

6.2.1 CLOUD COMPUTING SERVICE MODELS

Cloud computing provides services through the Internet to users and IT administrators. The evolution of cloud computing vanishes the installation of physical infrastructures, purchase of original software, platforms for operations, increased power consumption, need for cooling devices, etc. Cloud computing services can be classified into three types of service models that differ in cloud architecture and the level of abstraction. The three cloud services are as follows:

- **Platform as a service (PaaS)**: The platform as a service provides virtual resources for the development of web applications, which can be accessed through the Internet. The developed web application can be accessed by the authorized user without downloading or purchasing the required software. Instead, the required software can be utilized by the users through the utilization of cloud computing. The features provided by the PaaS are web application development, application hosting, resource security, scaling the usage level, etc. The major setback of the PaaS is the lack of portability of the service among the service providers and the users.
- **Software as a service (SaaS)**: The software as a service is a cloud model that enables the users to access web applications via Internet services. The SaaS enables users to access the services without any bifurcation in the types of operating systems, resources, processor frequency and type, etc.
- **Infrastructure as a service (IaaS)**: The infrastructure as a service provides users with access to virtual resources, and this service attracts many IT organizations to utilize this service and migrate from utilization of physical infrastructures to virtual infrastructures. This reduces the installation cost and maintenance efforts and cost and supports business continuity

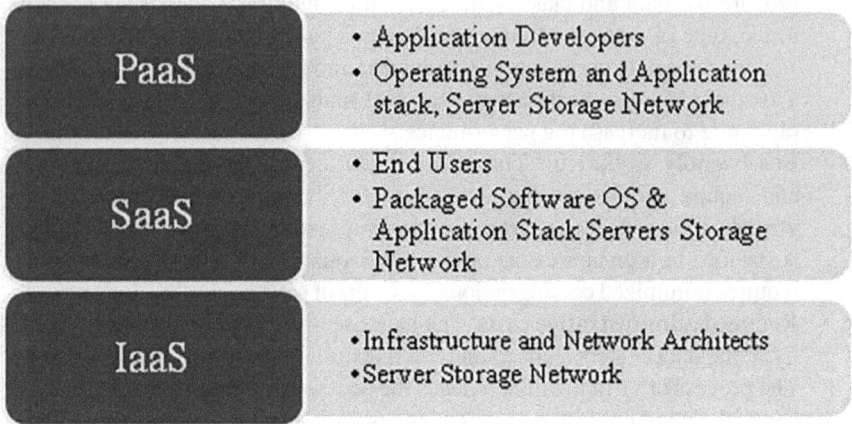

FIGURE 6.2 Cloud computing service models.

processes. The virtual clusters in the virtual infrastructure organize the software program required and processors for serving the user request, providing an effortless service to IT organizations.

Figure 6.2 narrates the concentrated details of the cloud computing service models, the beneficiary, and the way the users can be benefited.

6.2.2 CLOUD COMPUTING DEPLOYMENT MODELS

The cloud computing deployment model differs from the service model. The cloud computing hubs possess a cluster of server hubs and data storage devices to support the users with faster data transmission. To perform faster data transmission [10], it is necessary to position the data closer to the users. The data stored in the closer geographical area to the user can be delivered faster, and the cloud computing development model depends on the geographical area. The cloud computing deployment models can be classified into four types as follows:

- **Public cloud**: Public cloud is a deployment model in which the cloud services, resources, and infrastructures are made available to users who were classified under common public. The well-known examples for public cloud are Google, Amazon, etc., which are the cloud services accessed by the common public users all over the world. The data served to the users are classified based on their needs, constant search results, culture, etc.
- **Private cloud**: Private cloud [12] is a type of cloud deployment model in which the cloud resources will be accessed exclusively by a single organization. This type of cloud model is more secure when compared to the public cloud model, and the users have to undergo authentication verification process to prove their identity to access the resources in the private cloud. The well-known examples for private cloud are IBM and Microsoft, which are

the IT tycoons that offer services only to their customers and a selective group of users.

- **Hybrid cloud**: The hybrid cloud deployment model supports the business process to access the secured application from the cloud resources and is used by the common public. Since the protected resource of the IT organization is accessed by common public users, this deployment model is termed to be hybrid cloud model. The security level of the hybrid cloud is at medium level, which has a threat of certain vulnerabilities such as impersonation attack and password-guessing or brute-force attack.

- **Community cloud**: Community cloud is a deployment model in which the common cloud resource is shared by the community of IT organizations which are synchronized in their user policy, user data, etc. This type of deployment model reduces the infrastructure installation cost and maintenance costs. The well-known example for this type of cloud deployment model is the government website that can be accessed by various different types of departments.

Figure 6.3 depicts the shortest pictorial narration of access privilege and levels of security of the cloud computing deployment models.

6.2.3 CLOUD COMPUTING APPLICATIONS

The special characteristics of cloud computing, such as pay-per-use method, universal access to cloud resources, and on-demand services, attract many applications to be developed on the cloud computing resources, which were accessed by millions of users in 2020. As per the analysis report by the Grand View Research, the US cloud

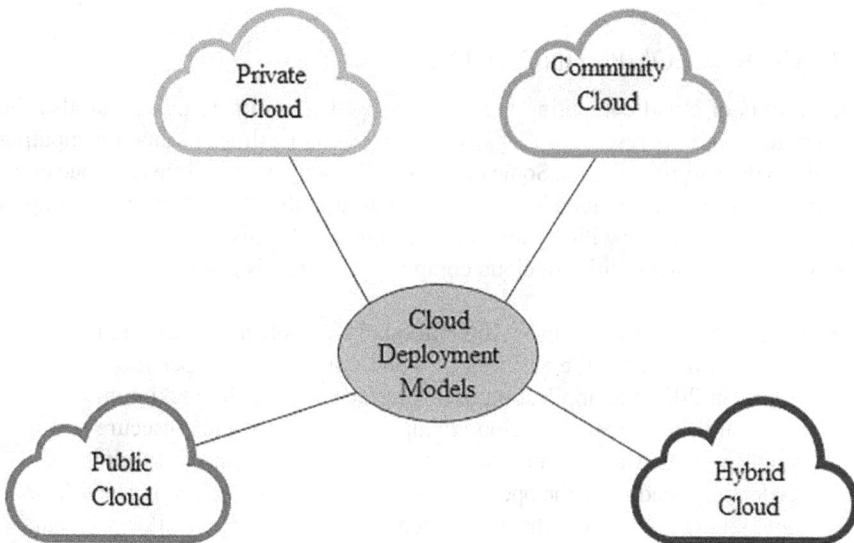

FIGURE 6.3 Cloud computing deployment models.

computing market size was over 101.4 billion US dollars. In this section, some of the well-known applications of cloud computing are listed.

- **E-learning:** Cloud computing has become a live trend in the teaching–learning process, and this has peaked during the pandemic period of COVID-19. Varieties of e-learning platforms and tools have evolved for teaching students so that they can gain knowledge without coming to the school premises. It is like an add-on facility along with the class room teaching method, where the faculty members, researchers, and students make use of cloud resources for their learning process. Some of the well-known e-learning applications that make use of cloud resources are Byju's and Vedanta.
- **Enterprise resource planning:** The development of business over the geographical area can be made possible with the evolution of cloud computing, and its notable application is called enterprise resource planning. This application makes use of private deployment models, where organizations make use of dedicated cloud resource such that the human resource management, accounts, and product management can be monitored from one place.
- **E-commerce:** e-commerce is the effective application of cloud computing where the leading IT tycoons market their products through the Internet. The organization has no necessity of installing or maintaining a huge database about the products, but can be maintained in the cloud resources. Some of the good examples for e-commerce are Amazon, Flipkart, etc.
- **Search engines and services:** The well-known search engine is the Google and its allied services, which is a public cloud deployment model in which the resources are commonly shared by the public users. Along with Google, Microsoft Azure and Amazon Web Services are the well-known search engines and services of cloud computing.

6.3   CLOUD COMPUTING VULNERABILITIES

The benefits of cloud computing attract not only the legitimate users, but also the hackers to invade its resources illegally and enjoy the facilities of cloud computing equally to the legitimate users. Some of the well-known vulnerabilities of cloud computing are illustrated here for a better understanding so that the mitigating techniques can be developed at par with the attacking techniques. In this section, a comprehensive list of top vulnerabilities of cloud computing security is given and narrated.

- **Insecure application user interfaces:** The application user interfaces are streamlined for the services of cloud computing. As per the Gartner report, in 2022 the application user interface will be the maximum factor of threat to the security of cloud computing services. This insecure application user interface occurs due to the poor verification of authentication credentials, leading to the open access of resources of cloud computing. The methodology to prevent the application user interface is to follow a strong level of authentication and to employ secure session layer (SSL) for data transmission.

- **Intellectual property:** The intellectual property rights (IPRs) are the most sensitive data that have to be preserved with care. The breach of IPR data leads to compromising or duplicating the product of an organization. The loss of IPR data leads to alteration of IPR data, deletion of data, and restricting the access to the legitimate user, which may lead to the complete loss of data from accessing it. To prevent the threat or loss of IPR data, it is advisable to implement data loss prevention program to safeguard the sensitive data in cloud computing resources.
- **Cloud storage misconfiguration:** Cloud storage is the golden asset for the hackers to invade in order to gain illegal access to the cloud resources. As per the Symantec report, during the year of 2018, around 70 million records were breached and accessed illegally from the cloud resources. This vulnerability is created due to the misconfiguration of cloud resource access and poor restrictions to the access of cloud resources. This can be prevented by restricting the access by configuring the access privileges properly.
- **Weak access policies:** The cloud resource must be accessed securely after the successful verification of authentication process. The poor security credentials framed by users can be broken using brute-force attack or password-guessing attack, leading to hackers to access the data in the form of legitimate users. To prevent this type of attack, strong authentication credentials can be framed and they have to be changed at a constant time interval.

Some of the reasons for the vulnerabilities of cloud computing security are given above, and the hackers launch certain types of attacks to invade the cloud computing resources. Some of the well-known attacks [9] with high level of threat weightage are listed and narrated as follows:

- **Brute-force attack**: The brute-force attack is the ancient methodology followed by the hackers in guessing the password in the trial-and-error method. The password can be cracked with minimal number of possibilities if the authentication credentials are weaker. The brute-force attack can be prevented by framing the authentication credentials stronger such that the brute-force attack consumes more time to be successful before which the session expires to save the cloud resources.
- **Man-in-the-middle attack**: The man-in-the-middle attack is a well-known attack in which the hacker tends to position himself in the middle of a legitimate user and the cloud server to steal the authentication credentials from the user and session ticket from the cloud server. This can be mitigated by improving the security credentials to a multi-level authentication mechanism in accessing the cloud resources.
- **Denial-of-service attack**: The denial-of-service attack is an attack intended to hide the availability of resources to the legal users. The hackers tend to send multiple request packets pretending to be a legitimate user. The cloud resource halts on the bulk amount of request from the hackers, leading to

TABLE 6.1

Survey on Cloud Computing Security

Title	Author	Conference/Journal	Mechanism	Limitation
Enhancing cloud computing security using AES algorithm	Abha Sachdev, Mohit Bhansali	*International Journal of Computer Applications*	Uses AES which is a block cipher algorithm and uses 128-, 192-, or 192-bit key [1]	It requires more computing cost as compared to DES
A hybrid approach for encrypting data on cloud to prevent DoS attacks	Navdeep Singh, Pankaj Deep Kaur	*International Journal of Database Theory and Application*	It uses AES followed by RSA algorithm—AES (128-, 192-, 256-bit key) and RSA (1,024-bit key)	Most complex system, need high-end processors [7], need more costly hardware. Time efficiency is less on slow hardware
Security in cloud computing using cryptographic algorithms	Shakeeba S. Khan, Prof. R.R. Tuteja	*International Journal of Innovative Research in Computer and Communication Engineering*	It uses DES followed by RSA algorithm—DES (64-bit key) and RSA (1,024-bit key)	DES is weaker than AES. Key size for DES is small. Can get access to system under brute-force attack [9]

non-availability of resources to the authorized users. This attack can be mitigated by scanning the network channel for malicious activities and preventing the malicious users to send furthermore requests.

Apart from the aforementioned attacks, certain more attacks such as impersonation attack and session hijacking attack also act as serious vulnerabilities for cloud computing. The survey on cloud computing security is depicted in Table 6.1.

6.4 CASE STUDIES ON ATTACK IN CLOUD COMPUTING

This section illustrates the real level of threats in the virtualization environment. This survey is conducted based on the worldwide virtualization attacks that have been reported in the last decade. At the end of the case study, the weightage of the threat can be analyzed. On analyzing the case studies, the victim applications were developed on their own platform and are implemented in the cloud environment. On surveying the well-known security attacks for the past decade, some of these were ransomware attack [3], Yahoo server attack, Mexican bank attack, Ascena Retail Group attack, security breach at dave.com, and data breach at the University of York. In July 2019, the ransomware launched by the Russian data hackers impacted the entire operation of the IT industry with a threat of losing its sensitive data to the hackers and in turn they demanded a huge ransom of bitcoins in exchange to release their data. Similarly, the Yahoo server attack is the major attack in which the user account details were breached and accessed by the hackers. The security attacks on all the aforementioned attacks and the entire attack history reveal that the principal way that permits the hacker to succeed is the poor authentication credentials and the poor security principles that were followed by the corresponding IT administrators. The constant process of auditing has to be performed to analyze the security principles and the data patterns in the network to train the incident response of the virtualization environment.

These applications were able to be accessed by users through the Internet, and the clients need to just install the application program or the application operating system through which the authenticated user can access the virtual resources online. To perform threat analysis, the data have been gathered from both the server end and the client end to determine the weightage of the threat in the virtualization environment.

$$TW_n = \frac{(TR_n * TF_n)}{\sum (TR_n * TF_n)} \tag{6.1}$$

where
TW—the weightage of threats
TR—the number of times the threat was repeated
TF—the number of victims affected by the threat
n—the number of users.

The weightage of the threat to the virtualization environment depends on two factors, namely the number of times the threat repeatedly occurred in a cloud computing environment and the number of victims who were affected by the threat.

6.5 SECURITY POLICIES AND RULES

The IT organizations, when migrating to the virtualization environment, has to consider the framing and maintenance of security policies and rules for preserving the privacy of the data stored in the virtual data centers. The common methods of implementing security policies are introduction of intrusion detection system and intrusion avoidance system in the network, which effectively track the data transferred in the network. The anomalous behavior of the data pattern is identified in the network, and the access is restricted for those adverse data patterns. The following are the security tips that have to be followed to maintain the security of the data stored in the virtual data centers.

- **Rapid update deployment**: The system admin has to be given with additional powers and privileges so that the admin can easily push the updates of the virtualization program that is installed in the virtual data centers.
- **Reviewing regulations**: To maintain proper security of the virtual data centers, the rules and regulations have to be reviewed at constant time intervals. A proper review has to be conducted on the rules and regulations to ensure the security of the virtual data centers.

- **Disabling unwanted hypervisor features**: The hypervisor is the software that acts as an interface between the host operating system and the guest operating system. The hypervisor governs the virtualization process. The hypervisor features are not completely utilized by the virtualization process, and it is necessary to disable the unwanted features of the hypervisor.
- **Active audition**: The network security policy, rules, and regulations have to be audited at constant time intervals, identifying the efforts of attacks such that regulations can be updated to mitigate the efforts of attackers in succeeding to obtaining the resources illegally.

The process of developing the virtualization security policies involves notable steps as follows:

- **Accountability and auditing**: Auditing and accountability are the roles and joint responsibilities of the network admin and users. The network admin has to perform network auditing at constant time intervals to install patch work if any flaws are identified in the network and virtual infrastructures. In turn, the users also hold equal responsibility for maintaining the security of the virtualization environment; that is, the user has to use their privileged role effectively and has to protect their account with strong login credentials.
- **Server role classification**: The server may hold multiple levels of processes such as file transfer server, domain server, e-mail server, database server, remote server, and project server. The security of the server can be maintained properly by classifying the roles of the server and partitioning it based on access, and giving privilege to the users accordingly.
- **Network service and configuration**: The virtualized network service is regularized by specifying the processes and roles of the network in access management list (AML), and it is essential to maintain the configuration of the network. The configuration creation, approval, validation, node inclusion and removal, etc. have to be managed properly and at constant frequency such that to maintain the undisrupted service.
- **Host security**: The host security deals with the selection of the location of the server, server monitoring process, and the access privilege factors for users to access the data resources.
- **Attack incident response**: This is the most essential process of security preservation, as the server may face attacks launched by the hackers and the network may trace the adverse behavior of packets in the channel. In this regard, the incident response policies have to be framed such that mentioning how the network has to respond for the incident identified and how the data have to be backed up without affecting the business continuity processes.

6.6   SECURITY SOLUTIONS FOR CLOUD COMPUTING

The case study and the analysis of the security threats to the virtualization environment and the effects after attack in the IT organizations make the users and the network administrators realize the necessity of the security solutions in the virtualization concept. The security solution for the virtualization lies on both the software and the hardware.

- **Defining security policies:** The IT organizations have to frame good security policies with rigid security rules, grounded instructions, and guidelines for preserving the security policies of the virtual infrastructure. The administrators have to collect relevant information such as the type of threat that attacks the cloud computing, severity of the threat, importance of the data, and compromised ports. It is insisted that the IT administrators seek cybersecurity professionals to perform an internal audit of the cloud computing for the identification of threats to the infrastructure, and based on the analysis report, an efficient security policy may be framed. Merely setting the security policy will not secure cloud computing, but the IT administrator has to review the policy and follow the policy at constant frequency of duration.
- **User awareness:** The security of the virtualization environment is not only the responsibility of the IT administrators but it is also a combined effort of IT administrators, organizational heads, and employees or users who work on or access the data from cloud computing. It is the responsibility of the IT administrators who have to explain the necessity of the security of sensitive data for the business continuity process, and above all, the users or the employees need to be aware of the necessity of maintaining the security policies. The users must be accountable for the data they access from the cloud computing. The users must be given privilege to access various levels of data, and it is not a good practice that all can access all sorts of data.
- **Utilization of cryptography techniques:** The network or IT administrators may incorporate a package of cryptographic techniques in the presentation layer so as to increase the level of security for the data transferred over the insecure network. As mentioned earlier, the users of virtual machine access the data from the server through the insecure channel of Internet. The Internet is a non-secure channel which can be obstructed easier by the hackers, and it is good to encrypt the data before transmitting and to decrypt the data at the authenticated receiver end so that a better level of security can be maintained. The choice of cryptographic algorithm relies on the IT administrators and the type of data stored in cloud computing, and due to the evolution of OSI model in the network, the network or IT administrator has an option for changing or migrating from one cryptographic algorithm to another as per the need.
- **Classification of information:** The data sets stored in the virtual server are not of single data type, but a collection of data types will be stored in the same virtual server. The data types may differ based on the organization; for example, a manufacturing company may collect and store different types of data such as client information and requirements [11], purchase data, manufacturing formula-related data, sales statistics, income and expenditure details, etc. It will be a good practice to classify these data with different levels of abstraction, instead of storing them as bulk data. The reason behind this classification is that, based on the classification, the user access privilege

can be provided to the user or the employees. Considering the aforementioned example, a sales team is able to access the sales-related information from the server so that the sales person will not be able to access the manufacturing-related information.

- **Defining business controls:** The process of defining business controls is a sort of access privilege that decides who can access and who can enjoy the edit privilege on the data stored in the cloud computing. The data stored in the virtual server are connected with the business continuity process, and hence it is necessary to define proper business controls. In the aforementioned example, the sales employee can access the data, whereas the sales manager enjoys the privilege of accessing the data and editing the data. The data editing privilege if given to the sales representatives will lead to leakage of data.
- **Network segregation:** The management process of a large network is not an easier task, and the network management will not be much effective. Hence, a good practice of segregating the large network into smaller networks linked by a common network head node increases the efficiency of network management and makes it an easier task for maintenance. The segregation of the network can be done based on the trust levels and data types and is done based on the logical networks as prescribed by organizations such as ISO and INU-N.

6.7   MITIGATING TECHNIQUES—A COMPARISON

The security threats can be mitigated by following the basic practices mentioned in the previous section. The risk of vulnerability can be mitigated by executing the following mitigating techniques.

- Framing effective policies, network management guidelines, and governing policies for virtual machine management life cycle.
- Controlling the creation of data, storage of data, and data access [13] privilege and defining the accessing tools.
- Employing the virtual patch detection techniques and the processes for updating, and application of security [14] configuration to cloud computing.

Figure 6.4 depicts the pictorial representation of mitigating techniques that may be followed to reduce the risk of vulnerabilities to cloud computing.

Figure 6.4 introduces five basic techniques for mitigating the vulnerability risk to cloud computing. The comparison of the mitigating techniques is explained below:

- **Firewall**: A suitable proxy firewall or application firewall installed in the IT organization restricts the inflow traffic to the organization and streamlines the data traffic and data access from the virtual server. This firewall filter monitors the network traffic in both ways, namely incoming data traffic and outgoing data traffic, such that the cloud computing may be protected effectively from the security attacks.

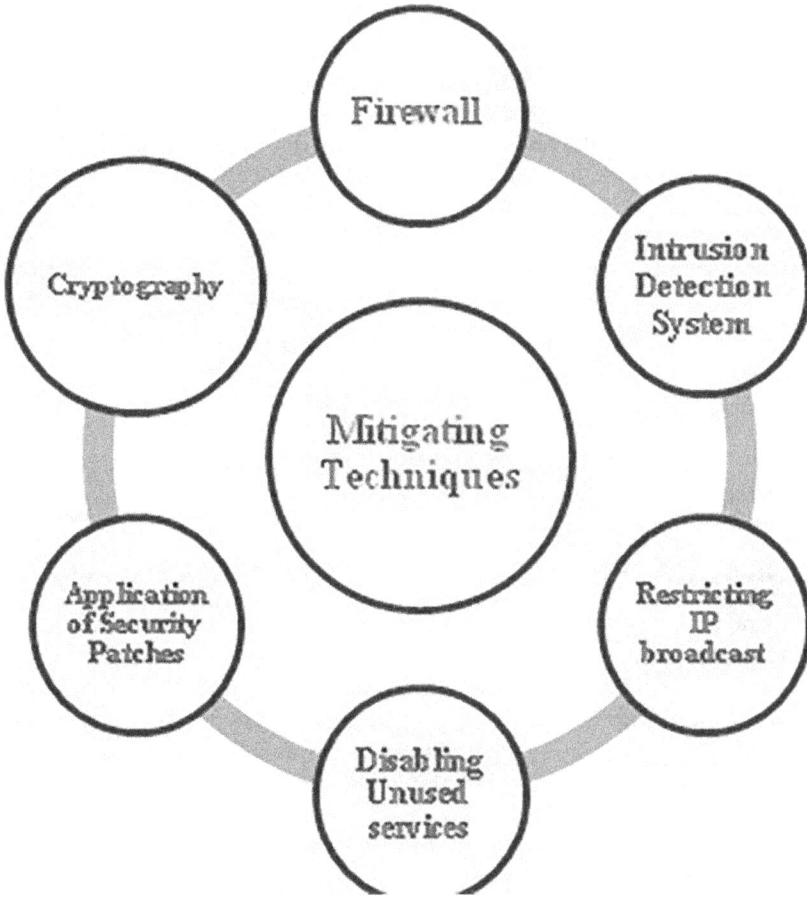

FIGURE 6.4 Techniques for mitigating the vulnerability risk to cloud computing.

- **Intrusion detection system (IDS)**: The IDS inspects the data traffic for anomalous behavior and gives alerts to the network administrator on detecting a suspicious activity or adverse behavior in the network. The IDS continuously scans the network traffic for malicious data and data breaching packets.
- **Restricting IP broadcast**: The IP broadcasting is a method in which a host system can amplify the incoming packets and will transmit the packet to all its cluster member nodes. In case of receiving a malicious program, the same will be amplified and forwarded to all the member nodes, leading to the security breach of entire network. Hence, it is a good practice to restrict the IP broadcasting process so that the network attacks and the vulnerability can be mitigated.
- **Disabling unused services**: The transmission protocols may be restricted to a single protocol, say Transmission Control Protocol, which follows a dedicated path for data transmission. On the contrary, the User Datagram

Protocol is a connectionless protocol, and when used in the network, the hackers use the same to launch their security attack. This may lead to bypassing the security policies and to breaching of security policies.

• **Application of security patches**: The host computer that is connected to the cloud computing has to be updated with a recent antivirus program or malware detection program so as to patch the security pitfall ports in the host system.

• **Cryptography**: The utilization of a proper cryptographic algorithm for data transfer and reception holds good in preserving the security of the sensitive data. The employment of a cryptographic algorithm mitigates most of the vulnerable attacks and preserves the cloud computing.

On comparing the mitigating techniques, each methodology holds equal importance in maintaining the security policy of cloud computing. However, the use of IDS, firewalls, and cryptographic algorithms plays a major role in mitigating the possibility of vulnerabilities and attacks in cloud computing.

6.8   ROLE OF CRYPTOGRAPHY IN SECURITY ENRICHMENT

Cryptography is a computing or mathematical tool for preserving the data by converting the plaintext to the ciphertext in the transmitter end, whereas the reverse process of conversion of ciphertext to plaintext is performed with the use of cryptographic key by the authorized user. Cryptography encompasses two process, namely encryption and decryption, in which the key plays a major role without which the cryptography is not possible. The key also holds the secrecy maintenance of the data, which is known only to the authorized users.

The cryptography assists the security preservation process by enabling the authorized sender and receiver to encrypt and decrypt the data. Figure 6.5 provides the pictorial representation of the cryptographic process, and from the diagram, it is clear that without possession or knowledge of key, one cannot decrypt the secret ciphertext. The cryptography ensures the security of the system and the integrity of the data. As per the analysis report of Global Encryption Trends in 2018 [12], around 43% of IT organizations were using strong cryptographic algorithms for preserving the integrity of their data stored in the virtual platform [15]. It has improved from 43% to 76% of IT organizations, which reveals that the organizations and IT administrators have realized the role of cryptography in securing cloud computing [17]. The cryptography does not only mean the standard algorithms [18–20] for encryption and decryption processes, but also involves the addition of secure certificates, inclusion of digital signatures, and performing hash algorithm to preserve the data from the unauthorized user access. The data and the virtual resources are vulnerable to the hackers of good knowledge and sophisticated attacking algorithms. The cryptography provides a high level of data integrity and confidentiality [16], and this has been proved by the security analysis performed in various types of networks. In the upcoming section, the notable cryptographic algorithms are discussed so that a better understanding of the cryptographic process can be achieved.

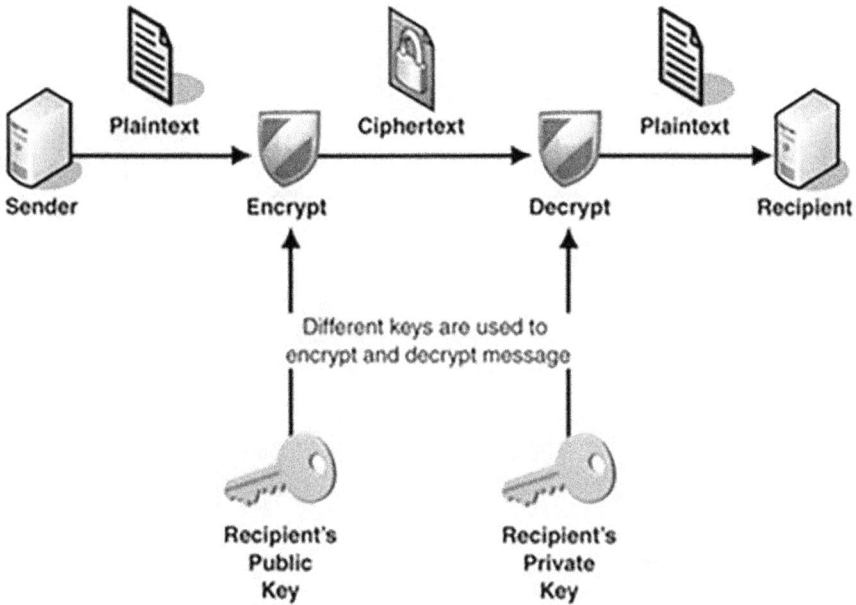

FIGURE 6.5   Role of cryptography in preserving data security.

6.9   COMPARISON OF CRYPTOGRAPHIC ALGORITHMS

Varieties of cryptographic algorithms have evolved from ancient cipher era until now with the aim of enriching the security level of the sensitive data. Some of the recent algorithms with notable features are listed below, and their features and functionalities are discussed.

- **Data Encryption Standard (DES)**: The DES is a type of symmetric key cryptography that accepts 64-bit plaintext as input and 48-bit data as key as shown in Figure 6.6. The output will be a 64-bit ciphertext which can be properly decrypted by an authorized user who possesses this 48-bit key.

  The DES has been proven to be more rigid against brute-force attack, as the algorithm is faster than the brute-force attack.
- **Triple DES**: The Triple DES algorithm was designed to overcome the disadvantages of the DES algorithm. The Triple DES algorithm [4] uses three secure keys, each of which is 56 bit long. Together, the key length is extended to 168 bits. The Triple DES holds good for hardware encryption for business process and IT organizations.
- **Advanced Encryption Standard (AES)**: The AES [5] is a symmetric key cryptographic algorithm with a secure key for encryption and decryption processes. The input bit size of the AES algorithm is comparatively more than the DES algorithm, with 128-bit input and 128-bit round keys as shown in Figure 6.7. The salient feature of the AES algorithm is the shift rows and

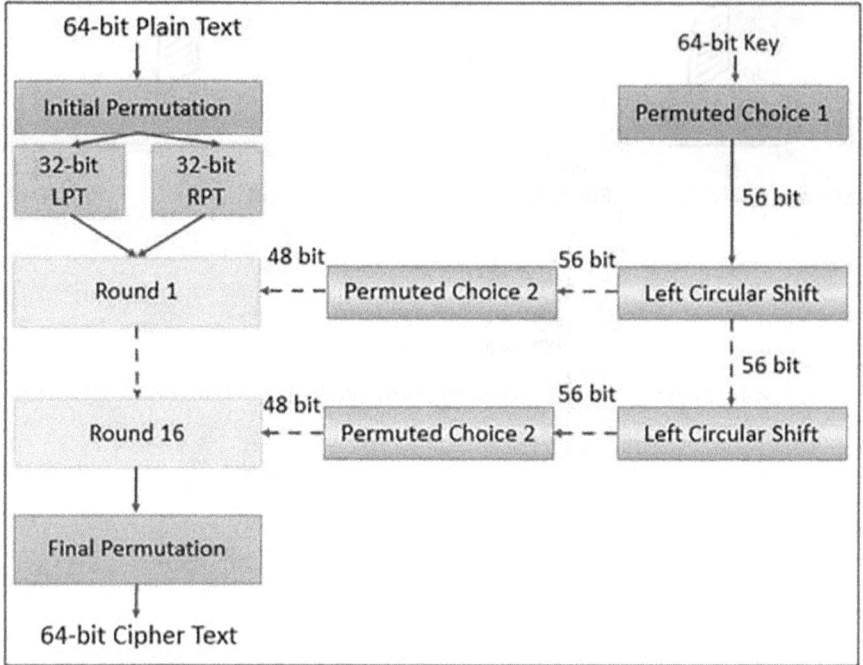

FIGURE 6.6 Data Encryption Standard (DES).

mix columns step, in which the plaintext is mixed with the round keys and is completely mixed to frame a non-readable ciphertext.

The AES algorithm is more rigid against brute-force attack, XSL attack, side channel attack, etc.

- **Rivest–Shamir–Adleman (RSA) Algorithm**: The RSA algorithm is an asymmetric key cryptographic algorithm implemented in modern computers for performing a secured data transfer, which is depicted in Figure 6.8. The RSA algorithm uses public key for encryption and private key for decryption process. The key selection is performed by the authorized users from a random selection of prime numbers p and q to determine the totient $\emptyset(n) = (p-1)(q-1)$. From this totient, the congruent relation is determined for encryption process.

  The binary bit example is illustrated in Table 6.2.

- **Blowfish**: The Blowfish [6] is a cryptographic algorithm, which was designed in view of replacing the DES algorithm to overcome its disadvantages. The well-known feature of the Blowfish algorithm is its security level and its computing speed. The free availability of the Blowfish algorithm attracts many users to install this algorithm in their organization.

- **Twofish**: The Twofish algorithm [7] is a successor to the Blowfish algorithm, and it uses a key of 256 bit length and follows a symmetric method of encryption process. Similar to the Blowfish algorithm, the Twofish algorithm is well known for its computation speed and is freeware similar to the Blowfish algorithm.

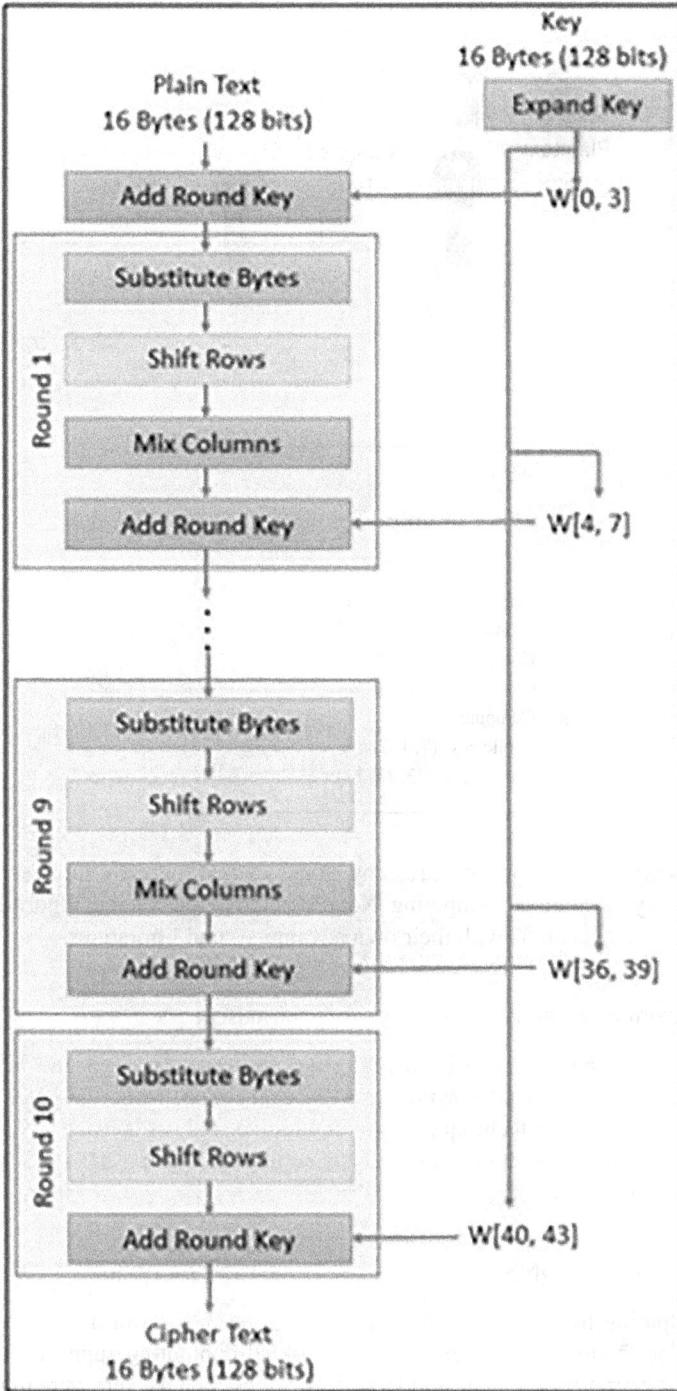

FIGURE 6.7 Advanced Encryption Standard (AES).

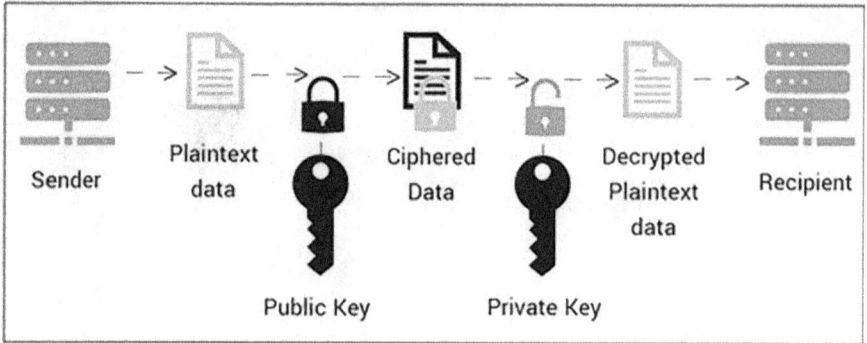

FIGURE 6.8   RSA algorithm.

TABLE 6.2
RSA Example

RSA Example:

Select primes: $p = 17$; $q = 11$
Compute $n = p * q = 187$
Compute totient: 160
Generator "e" = 7
Compute "d" = 23
Public key: (7, 187)
Private key: (23, 17, 11)

The above-explained algorithms are ruby stones of cryptography in preserving the security policy of the cloud computing. Not restricted to the count of algorithm, there exist numerous algorithms with their own advantages and limitations.

6.9.1   CRYPTOGRAPHIC ALGORITHMS—A COMPARISON

The comparative analysis had been performed between the DES and RSA algorithms. The DES incorporates symmetric key encryption, while the latter involves public key cryptography technique. The comparative analysis is listed in Table 6.3.

The graphical representation of analysis results comparing the DES and RSA algorithms is portrayed in Figure 6.9.

6.10   CONCLUSIONS

Cloud computing has been the computing revolution in the world for the past two decades. The features and characteristics of cloud computing support and render various applications to the users and IT organizations. The hackers were highly considered due to the advances in the hacking technology to access and enjoy the cloud computing resources. Various researchers have actively been involved in finding

TABLE 6.3

Comparative Analysis of DES and RSA Algorithms

Parameters	DES	RSA
Key length	56 bits	>1,024 bits
Cipher block	64 bits	Minimum 512 bits
Algorithm	Symmetric	Public key cryptography
Encryption duration	Medium	Slow
Decryption duration	Medium	Slow
Security level	Moderate	Moderate
Rounds	16	One
Simulation speed	Fast	Fast

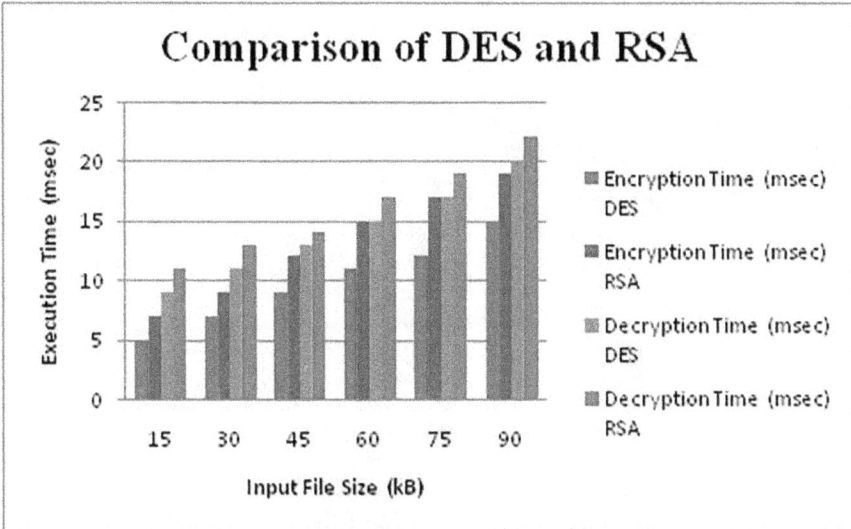

FIGURE 6.9   Comparison of DES and RSA.

different solutions for enhancing the security in the cloud computing paradigm. The security aspect of cloud computing relies only on the authentication credentials, and in future, this can be enhanced with latest technologies such as blockchain technology, machine learning, deep learning, and artificial intelligence, which make the cloud resource to be self-smart in authentication verification process.

REFERENCES

1. D. Wang, K. Zhang, M. Qiang, X. Jia, and Y. Chen (2020). Computer-assisted pre-operative planning improves the learning curve of PFNA-II in the treatment of inter-trochanteric femoral fractures. *BMC Musculoskeletal Disorders* 21. doi: 10.1186/s12891-020-3048-4.

2. R. Bhushan and D. Yadav (2020). Verification of virtual machine architecture in a hypervisor through model checking. *Procedia Computer Science* 167, 67–74. doi: 10.1016/j.procs.2020.03.183.

3. V. Petrucci, O. Loques, and D. Mossé (2010). A dynamic optimization model for power and performance management of virtualized clusters. *Proceedings of the e-Energy 2010– 1st International Conference on Energy-Efficient Computing and Networking*, 225–233. doi: 10.1145/1791314.1791350.

4. S. Davies, R. Macfarlane, and W. Buchanan (2020). Evaluation of live forensic techniques in ransomware attack mitigation. *Digital Investigation* 33. doi: 10.1016/j.fsidi. 2020.300979.

5. A. Vuppala, R. Roshan, S. Nawaz, and R. Jvr (2020). An efficient optimization and secured triple data encryption standard using enhanced key scheduling algorithm. *Procedia Computer Science* 171, 1054–1063. doi: 10.1016/j.procs.2020.04.113.

6. M. Masoumi (2019). A highly efficient and secure hardware implementation of the advanced encryption standard. *Journal of Information Security and Applications* 48. doi: 10.1016/j.jisa.2019.102371.

7. P. Patel, R. Patel, and N. Patel (2016). Integrated ECC and blowfish for smartphone security. *Procedia Computer Science* 78, 210–216. doi: 10.1016/j.procs.2016.02.035.

8. S. Kareem and A. M. Rahma (2020). A novel approach for the development of the Twofish algorithm based on multi-level key space. *Journal of Information Security and Applications* 50, 102410. doi: 10.1016/j.jisa.2019.102410.

9. R. Shyamala and D. Prabakaran (2018). A survey on security issues and solutions in virtual private network. *International Journal of Pure and Applied Mathematics* 119(14), 1183–1192.

10. D. R. Bharadwaj, A. Bhattacharya, and M. Chakkaravarthy (2018). Cloud threat defense: A threat protection and security compliance solution, *In 2018 IEEE International Conference on Cloud Computing in Emerging Markets (CCEM)*, Bangalore, India, pp. 95–99. doi: 10.1109/CCEM.2018.00024.

11. J. Koo, Y. Kim, and S. Lee (2019). Security requirements for cloud-based C4I security architecture, *In 2019 International Conference on Platform Technology and Service (PlatCon)*, Jeju, Korea (South), pp. 1–4. doi: 10.1109/PlatCon.2019.8668963.

12. L. Qing, Z. Boyu, W. Jinhua, and L. Qinqian (2018). Research on key technology of network security situation awareness of private cloud in enterprises, *In 2018 IEEE 3rd International Conference on Cloud Computing and Big Data Analysis (ICCCBDA)*, Chengdu, China, pp. 462–466. doi: 10.1109/ICCCBDA.2018.8386560.

13. A. Markandey, P. Dhamdhere, and Y. Gajmal (2018). Data access security in cloud computing: a review, *In 2018 International Conference on Computing, Power and Communication Technologies (GUCON)*, Greater Noida, India, pp. 633–636. doi: 10.1109/ GUCON.2018.8675033.

14. X. Sun (2018). Critical security issues in cloud computing: a survey, *In 2018 IEEE 4th International Conference on Big Data Security on Cloud (BigDataSecurity), IEEE International Conference on High Performance and Smart Computing, (HPSC) and IEEE International Conference on Intelligent Data and Security (IDS)*, Omaha, NE, pp. 216–221. doi: 10.1109/BDS/HPSC/IDS18.2018.00053.

15. Z. Li, L. Liu, Y. Zhang, and B. Liu (2018). Learning and predicting method of security state of cloud platform based on improved hidden Markov model, *In 2018 3rd International Conference on Smart City and Systems Engineering (ICSCSE)*, Xiamen, China, pp. 600–605. doi: 10.1109/ICSCSE.2018.00128.

16. Y. Wu, Y. Lyu, and Y. Shi (2019). Cloud storage security assessment through equilibrium analysis. *Tsinghua Science and Technology*, 24(6), 738–749. doi: 10.26599/TST. 2018.9010127.

17. K. Suresh Kumar, A. S. Radha Mani, S. Sundaresan, and T. Ananth Kumar (2021). Modeling of VANET for future generation transportation system through edge/fog/cloud computing powered by 6G. In: Gurinder Singh, Vishal Jain, Jyotir Moy Chatterjee, Loveleen Gaur (Eds) *Cloud and IoT-Based Vehicular Ad Hoc Networks*, pp. 105–124. Hoboken, NJ: John Wiley & Sons.

18. J. Singh, T. Pasquier, J. Bacon, H. Ko and D. Eyers (2016). Twenty security considerations for cloud-supported internet of things. *IEEE Internet of Things Journal* 3(3), 269–284. doi: 10.1109/JIOT.2015.2460333.

19. Y. Yang, R. Liu, Y. Chen, T. Li and Y. Tang (2018). Normal cloud model-based algorithm for multi-attribute trusted cloud service selection. *IEEE Access* 6, 37644–37652. doi: 10.1109/ACCESS.2018.2850050.

20. K. Salah, M. Hammoud, and S. Zeadally (2015). Teaching cybersecurity using the cloud. *IEEE Transactions on Learning Technologies* 8(4), 383–392. doi: 10.1109/TLT.2015.2424692.

21. Gartner Analysis Report: https://www.gartner.com/technology/research/.

22. Spicework Research Report: https://www.spiceworks.com/research/.

23. Global Encryption Trends Study: https://www.entrust.com/digital-security/c/global-encryption-trends-study.

7 Secure Data Storage and Retrieval Operations Using Attribute-Based Encryption for Mobile Cloud Computing

S. Usharani, K. Dhanalakshmi, P. Manju Bala,
R. Rajmohan, and S. Jayalakshmi
IFET College of Engineering

CONTENTS

DOI: 10.1201/9781003219880-7

7.1   INTRODUCTION

Recently, many cloud services are available in the market, and they provide two primary services: cloud storage and computation. Even though cloud storage and computation provide many advantages to the users, they are still hesitant to move to the cloud services because the data in the cloud storage are maintained in public clouds, which are accessible to other cloud customers who can be malicious attackers. The customer does not trust the service providers. Moreover, there is a possible risk that the multi-tenant architecture may expose the data to a potential competitor or malicious attacker, which may compromise the data security.

When they choose to secure the information remotely in cloud storage, data confidentiality and integrity are the two main concerns of the users. Encryption is enabled to secure sensitive information stored in multiple storages to safeguard user data. Ciphertext-policy attribute-based encryption (CP-ABE) remains a new method intended for managing cloud access control, effectively enabling key management and authentication. A descriptive string called attribute is allocated or related to such a user in the CP-ABE construction. One user can have attribute values in the same way that one attribute can be assigned to multiple users. This allows the service provider to a data access policy that comprises the user's multiple attributes through logical operators such as "OR" and "AND." User attributes must match the access policy to access the encrypted message. This distinctive feature makes the CP-ABE solution suitable for a cloud storage system that requires a large user base from diverse backgrounds to have an efficient data access mechanism.

With the exponential advancement of technology, models of mobile cloud systems are the focus of the future, wherever mobile devices are cast off as nodes for the collection and dispensation of data. This requires researchers from the cloud service to build a reliable cloud computing architecture that includes lightweight mobile devices with fewer computing resources. A new issue is incorporating lightweight connected networks such as cell phones and tablets in the public cloud with the CP-ABE-allowed cloud service; this problem arises because CP-ABE needs high computational resources on doing encryption and authentication. The solution to these problems is automating the cryptographic operations to the cloud server without exposing the user's data material. The proposed solution SC-CP-ABE focuses on this issue and provides an efficient and effective way to outsource the data to the cloud without revealing its content. Another challenge with the mobile cloud models is sharing data to the frequently changing data groups. For example, when file access to a particular user is revoked, the user cannot view any potential system changes, and the regional backup of that file will become redundant. The modified file must be encrypted with the latest encryption key to do this. Another concern is how to do this encryption and decryption. For example, when there is a change in a particular datum, it requires the data to be downloaded for decryption and re-encrypted when uploading the data back to the server. This frequent data download and upload will impose significant overhead on lightweight wireless devices. To prevent this, we need a stable and robust public cloud system to handle overhead storage and communication. This chapter suggests a solution to overcome this problem mentioned above, which comprises two key components: First, secrecy-conserving CP-ABE (SC-CP-ABE), which enables the user to farm out the dense cryptographic operations of the CP-ABE to cloud server without exposing the data content or cryptic key; second, attribute-based data retrieval (ABDR), which enables the optimal cloud data access and management system. Using ABDR, the user attributes are maintained in a tree-like hierarchy that reduces the cost of membership revocation. It also balances the message and the computation overhead for the mobile nodes and the storage service providers.

7.2 CLOUD COMPUTING

In the world of data management, cloud computation is an emerging technology. Cloud computing uses computational services that are distributed as a utility over a network such as Internet. Cloud computing defines highly scalable services supplied as an external commodity on a pay-as-you-use basis over the Internet. Cloud computing's name derives from the cloud-shaped icon used to elucidate a web-connected remote resource. It can be described as follows. Most of them don't think where the power is produced or transported. They simply use the electricity without any knowledge.

At the end of every month, they have to pay for the electricity consumed. The concept of motivating cloud computing is very similar. People may easily access and use the resources such as computing power, storage, or custom-built production environments. They don't have to worry about how they operate internally. The primary principle for cloud technology is that consumers use what they need and are charged for what and how much they use.

7.3 MOBILE CLOUD COMPUTING

Mobile cloud computing (MCC) exploits elastic cloud structure across heterogeneous cellular networks to expand capacities for mobile devices that are fundamentally resource-constrained. Thus, MCC consists of a three-tier architecture composed of mobile devices, multiple cellular networks, and cloud-based resources to provide mobile customers with cloud-based applications. Each layer shows its security and privacy concerns. In MCC, execution time and energy usage are dramatically increased by outsourcing resource-intensive activities from the portable device to the cloud service network. However, leveraging remote capital poses many security challenges, such as faith, security, availability, or confidentiality. Also, connectivity across heterogeneous cellular networks consumes more energy than in wired communication, so there is a need to reduce the use of mobile system capital to provide mobile end users with sustainable on-demand services. User verification and user permission are two of the most critical protection problems in mobile cloud computing. In this chapter, we concentrate on user authorization of outsourced data processed on the cloud storage network. Our solution should be resource-friendly for mobile devices while enforcing data protection and confidentiality.

7.4 SYSTEM ARCHITECTURE

Figure 7.1 illustrates the system framework of MCC. On a mobile device sending a route request to the server, the server sends the connection request to the cloud server. The cloud server will check the data availability, authentication, and other services. If the client has the required authorization rights to use the data, then the data will be provided from the backend services to a mobile client.

7.4.1 SERVICES NECESSARY FOR MOBILE CLIENT

A discussion of some of the facilities required by a mobile cloud client is given below:
- **Synchronization**: It monitors and synchronizes state changes, if any, to the mobile or its program.
- **Push**: Push handles every state change from the cloud server by pressing.
- **Offline application**: The communication between services such as synchronization and push is controlled and developed.
- **Network**: It quickly maintains proper coordination and manages the user's communication channel to accept push notifications.
- **Database**: Databases handle the local data storage for smartphone applications.

7.4.2 SERVICES OFFERED BY MOBILE SERVER

A discussion of some of the facilities offered by mobile cloud servers is presented below:

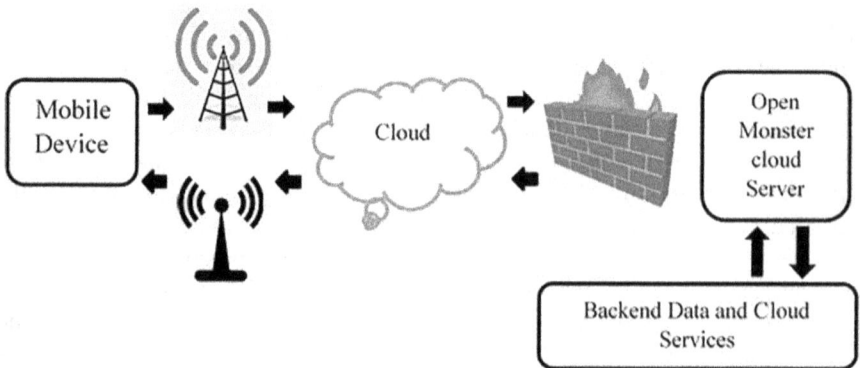

FIGURE 7.1 System architecture for mobile cloud computing.

- Synchronization: It synchronizes device-side application changes with the existing data position. The data on the backend must also be mobilized.
- Push: It tracks backend facts sources for notifications. Smartphones are alerted about this after updates are found.
- Secure socket data service: This service must have a simple socket server or an SSL-based socket server, or both, based on security specifications.
- Security: This service offers authentication and authorization services to allow the system to be accessed by mobile devices connected to cloud servers.

7.5 LIMITATIONS AND ISSUES IN MCC

Although the processing capability of portable devices has a limit, cloud is incredibly efficient to execute computations, so many challenges exist to demonstrate how to balance the discrepancies between these two. There are several difficulties with the application of mobile cloud computing. These problems may be related to insufficient infrastructure, network-related issues, smartphone user protection issues, and cloud issues. Some specific problems are explained below.

7.5.1 RESOURCE LIMITATIONS

Having limited mobile storage infrastructure makes it challenging to use cloud storage on personal devices. Limited processing capacity, limited battery, and poor-quality monitor are simple drawbacks due to limited resources.

7.5.2 NETWORK ISSUES

All computations in MCC are done on the network. There are also several network-related problems, such as bandwidth, latency, compatibility, and heterogeneity.

7.5.3 SECURITY ISSUES

Like a personal computer, most handheld devices have about the same functionality. Thus, applications still have to face a range of security and privacy concerns. Services are now provided on clouds to resolve this question of vulnerability identification, but this still needs to meet several difficulties. Any protection problems include device stability, smartphone user privacy, and cloud service security. These attacks influence the information collected on the cloud. The credibility of the information is significant to the owner. If any unauthorized person makes modifications to another person's data, the data's credibility can be impaired. Any individual may damage another person after discovering sensitive information from another person. Data security is also a problem for the owner of the data. User authentication is often necessary to verify who the file originator was. There are so many challenges to security, such as malware, ransomware, and Trojan horses. Privacy issues are triggered by the use of the global positioning system on mobile platforms [16].

- **Security issues in a mobile terminal:** It is an enterprise planning system that allows, anywhere, at any time, wireless Internet access. It also encourages tech and third-party personalization. Security concerns are, therefore, significant in mobile terminals, and they address these with respect to ransomware, system failures, and other points of view as well.
  - **Malware:** Malware retains access to users' personal information when accessed and downloaded immediately, which is inaccessible to users. Many anti-malware applications have been developed, but essential computing resources are challenging to attain due to restricted resources and mobile terminal power. So, solutions are required for the identification and prevention of malware in mobile terminals.
  - **Software vulnerabilities:** FTP is used to move the user name and password to the network, and they are saved in plaintext format. This makes it easier to hack cell phones illegally from machines on the same network, meaning that sensitive information is not safeguarded.
    As in the operating code, coding vulnerabilities occur, leading to the loss of cell phones by criminals in certain situations.
- **Security issues in mobile network:** Mobile users can use the network in many ways, such as calling networks, transmitting short message services (SMS), and other Internet assistance. Smartphones can also be reached via Wi-Fi and the Bluetooth network. Thus, these types of entry result in security threats and malware attacks.
- **Mobile cloud protection problems:** The security of the mobile cloud is discussed in terms of two problems, namely reliability of the network and protection of data and privacy.
  - **Reliability of a network:** High storage of information resources are provided by cloud, which lead to threat and possibility of attacks. These risks may be from malware outside, cloud users, or employees. The hackers aim to disrupt cloud suppliers. Denial of service, for instance, stops the cloud service by eliminating the network available.
  - **Privacy and data protection:** User data rights and maintenance are housed in different sites, and customers often do not know the precise location of the infrastructure where their information is stored. In the mobile cloud storage world, data security and privacy are therefore of considerable concern.

7.6   RELATED WORKS

The current work relating to systems includes (i) encryption centered on attributes and (ii) regulation of computational control to unreliable information. As a blurry variant of IBE, where an entity is seen as a set of improving the rights, attribute-based encryption was primarily proposed [1,2]. So far, two prior versions of attribute-based encryption have been suggested, i.e., key-policy attribute-based encryption (KP-ABE) and ciphertext-policy attribute-based encryption (CP-ABE). Each ciphertext in KP-ABE is paired with a group of features, and each user's private key is encoded with an information system. Decryption is permitted only if the ciphertext parameters comply with the user's access policy for private keys [3,4]. In CP-ABE, each user has a group of parameters associated with the customer's private key, and an information system encrypts each ciphertext [5]. To decipher the message, the customer's private key features must comply with the information system.

CP-ABE is more attractive because the function-based access control paradigm is conceptually closer to it. In both the cryptography community and the networking community, cryptographic access control over untrusted storage is studied [6,8]. Compared to traditional one-to-one encryption, broadcast encryption is very secure and easy. Current broadcast encryption will identify the compromises between key storage and overhead ciphertext storage as (i) constant ciphertext for the maximum amount of recipients, regular public and/or secret key; (ii) linear ciphertext for the number of receivers removed, constant (or logarithm) public and/or private key; and (iii) sub-regular ciphertext, public and/or secret key sub-regular code [9–12]. In this chapter, a novel ABDR scheme edifice to resolve the lack of all current mechanisms. ABDR enables any random number of receivers with much lower storage and communication complexity. In the networking community, various object storage systems have been proposed to protect data via untrusted storage. Specifically, the authors introduced a distributed storage scheme where consumers automate encryption to a semi-trusted re-encryption server. However, the data's privacy would be violated if the network collides with unauthorized access [13–15]. Our suggested SC-CP-ABE is secure relative to this system, while providers of services and malicious users collaborate. It is a cloud computing security framework based on CP-ABE. The approach allows consumers to share the unique secret key to our job to the cloud, although our result only sends blind secret keys. In particular, our results particularly deliberate mobile cloud environs and their exertion.

7.7   PROPOSED SCHEME

A mechanism/scheme for providing safe data retrieval and storage in MCC is proposed here. This framework utilizes the concept of the hashing algorithm along with other encryption and decryption methods to include better protection for the data stored on the cloud platform. A trusted third-party auditor (TPA) is a trustworthy person. TPA tests, on behalf of the data holders, the data integrity processed in the mobile cloud. To verify the data integrity, TPA verifies the hash and post. The TPA offers credibility checking, which eliminates a lot of the smartphone user's work. In this method, the data owner has two keys, namely a private key and a public key. The file/message is ciphered by twice a time. First, the message should be encrypted with the user's private key, and second, the same encrypted file is once again encrypted with the TPA public key. This then grants the details of smartphone user's secrecy.

A holistic, stable portable cloud management architecture involves two key elements:
1. A secrecy-conserving CP-ABE (SC-CP-ABE) scheme.
2. A method of ABDR that maintains fuzzy knowledge optimality.

Using SC-CP-ABE, consumers can easily automate computationally CP-ABE cryptographic systems to the cloud, deprived of exchanging data materials and secret keys. In this method, the compact and reserve-restricted computers can connect and work with information stored in a cloud data warehouse. The ABDR scheme achieves versatile and quite good data security using public cloud services. Centered on ABDR, users' characteristics are arranged in a deliberately designed hierarchy to minimize the cost of revocation of membership. Moreover, ABDR is sufficient to combine overhead networking and storage for mobile computing, minimizing the risk of data processing operations for both cloud computing nodes and database service providers (such as uploads, upgrades, and so on).

7.8   SYSTEM AND MODELS

In this section, we are going to see about notations and system design.

7.8.1   NOTATIONS

TABLE 7.1
Notations Used in This Chapter

Abbreviation	Expansion
DO	Data owner
TPA	Third-party auditor
CSP	Cloud service provider
ESP	Encryption service provider
T	Access policy tree
SSP	Storage service provider
DSP	Decryption service provider

7.8.2   SYSTEM DESIGN

In our suggested process, the data owner is denoted as DO. A portable handheld device that can order and/or collect data in the cloud can be a DO. By using our suggested SC-CP-ABE system, the data are safe. There are several data recipients, other than DO. Subscriptions to the data are maintained by the Data Owner. The obtainable computer model has the following characteristics: The file or data before being sent to the storage service provider should be encrypted.

1. The encryption service provider offers an encryption facility to the DO underprivileged of significant the unique key for data encryption (KDE).
2. Without knowing the files' content, the decryption service provider provides the clients with a decryption service.
3. The data quality will not be exposed except through ESP, DSP, and SSP collusion.

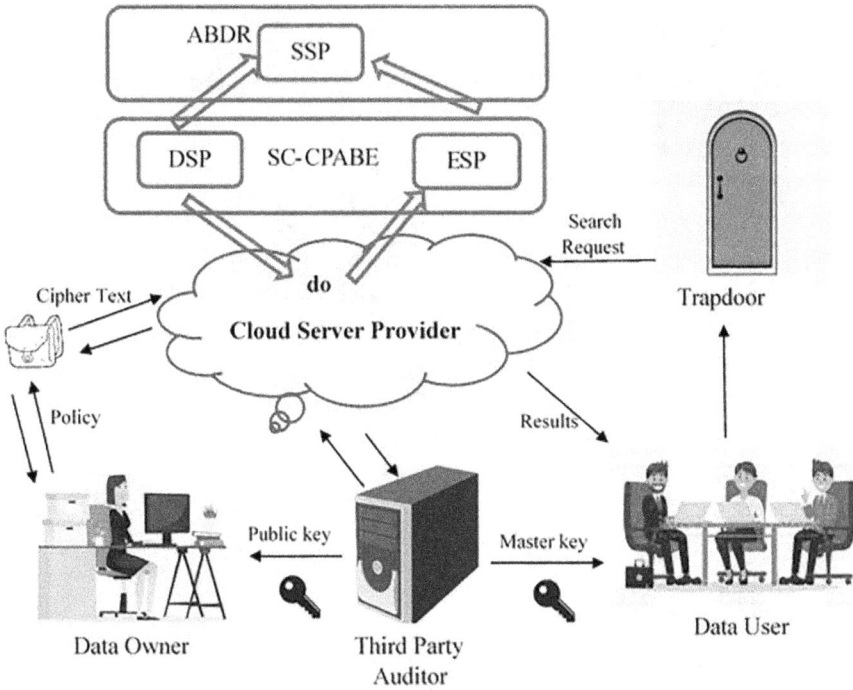

FIGURE 7.2   System design.

The storage, encryption, and decryption service providers constitute the proposed scheme's central modules, as shown in Figure 7.2. The encryption service provider offers SC-CP-ABE facilities, and the decryption service provider and storage service provider provide storage services. The cloud is semi-loyal, in which the cloud provides only computing and storage services with machine security assistance, although the information is invisible to the cloud. More authoritative PCs and mobiles can serve as message proxies for devices that, in particular, collect data.

Along with extracting each data file's keywords, DO generates a secure-searchable keyword index until outsourcing them to the CSP with such an attribute-based access policy. Many facilities, such as data collection, computing, and search, are the responsibility of the CSP. If a DU wishes to send a search query over encrypted files using its private key, it will create a search trapdoor according to the relevant keyword and apply it to the CSP. The CSP tries to verify whether the specified search keyword fits the index after obtaining the trapdoor, despite gaining knowledge of the encrypted data and the searching keyword. Finally, the CSP provides the relevant results to the DU, assuming the trapdoor data consumer attributes meet the secure-searchable index access policies. The search trapdoor fits the index of keywords. Furthermore, a TPA is responsible for creating and circulating public keys and master keys. Figure 7.3 demonstrates the communication between DO, TPA, and providers of cloud services.

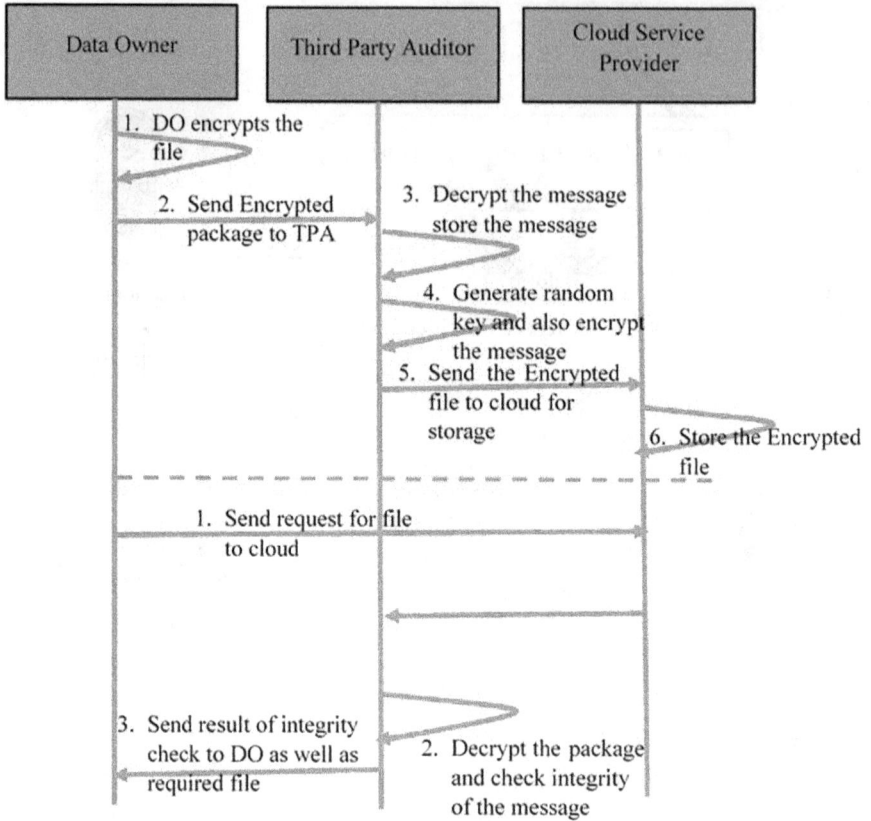

FIGURE 7.3   Interaction among DO, CSP, and TPA.

7.9   ATTRIBUTE-BASED ENCRYPTION

Attribute-based encryption was proposed as a workable solution of taking extra in outsourced data on cloud computing environments. It already has a versatile encryption strategy and good control of access that has effectively accelerated its acceptance rate in outsourced data management schemes that are outsourced. The DO in ABE encrypts the data in compliance with an information system and can only be decrypted by the user's key attributes that match the cipher document's attributes. Two major types exist in ABE schemes, namely KP-ABE and CP-ABE, based on whether the user's hidden key or ciphertext determines the access policy. Multi-authority ABE was suggested to allow consistency in access management across several entities or domains and enhance efficiency, where separate attribute authorities individually control attributes and where both security and responsibility are spread across multiple attribute authorities. The literature on multi-authority attribute-based schemes has presented numerous challenges. However, the latest work on multi-authority ABE still displays the critical bond constraint if one of the authorities requires ample rights to read and decode user messages. Also, because

ABE operations typically require high overhead computation and communication, the proposed scheme should carry minimum computation and communication costs.

7.10   SECRECY-CONSERVING CP-ABE

7.10.1   Construction

The algorithm for SC-CP-ABE is constructed for secrecy conserving. In SC-CP-ABE, without exposing their data quality and hidden keys, intensive CP-ABE encryption and decryption computation is a farm out to effective cloud service providers by DOs. The knowledge of SC-CP-ABE is to farm out the service providers to an extreme and non-essential portion of the algorithm for encryption and decryption while retaining sensitive secrets. However, in contrast to conventional CP-ABE, the outsourcing of complete computation does not minimize the level of protection is performed locally. CP-encryption ABE's complexity rises linearly on the SCP-encryption. The complexity of ABE increases with the size of the access control in linear terms. During encryption, a master key is embedded in the ciphertext in a recursive method. As per the tree of access policy, the code is divided into all the sub-trees of the present root policy at each point of the information system. The level of security is, therefore, different from the access control tree. In other words, even though the ESP has secrets from many other but not all sections of the access control tree, the master secret is still legally secure data, provided that there is at most one information undisclosed to the ESP. Thus, by providing a lesser amount of confidential data in the vicinity, they can easily outsource many of the authentication problems to the encryption service provider. However, the CP-ABE decryption algorithm is time-intensive for decryption because bilinear pairing operations via ciphertext and secret key are a rigorous computer process. The CP-ABE decryption algorithm takes time. SC-CP-ABE overcomes this programming problem by distracting the private key safely and exporting to the DSP the costly pairing functions. Again, the subcontracting would not show the data material of the ciphertext to the DSP. This is since the decryptors do the final decipher stage.

7.10.2   Background Information

Here, some preliminary approximate information of our architecture is provided in this chapter. The suggested SC-CP-ABE is intended to use both piecewise linear matching and secret sharing frameworks, which are briefly discussed below.

1. **Bilinear pairing**: A bilinear map function $f : M_0 \times M_1$ is a pairing of any two multiplicative cyclic groups of huge prime order p are M_0 and M_1. On both $M-0$ and $M-1$, the discrete logarithm problem is complicated. The following properties apply to pairing:
   - Bilinearity:

$$f\left(L^x, N^y\right) = f(L,N)^{xy}, \forall\, L, N \in M_0, \forall a, b \in \mathbb{Z}_p^*.$$

- Nondegeneracy:

$$f(m,m) \neq 1 \quad \text{where } m \text{ is the generator of } M_0$$

- Computability:
  An effective pairing estimation algorithm exists.

2. Hidden sharing: (s, b)

Hidden sharing is being done to split a hidden into b parts, and some s shares will rebuild the hidden, although no data about the top-secret can be exposed by combining fewer sharing than s shares. As Shamir et al. [16] launched in a polynomial of $s-1$ degree, every point of s on the polynomial is made to rebuild the polynomial. The Large range coefficient $\Delta_{j,Q}$ for $j \in \mathbb{Z}_p$ and a set, Q, of features in $\mathbb{Z}_p$ is:

$$\Delta_{j,Q}(a) = \prod_{(i \in Q, i \neq j)} \frac{a-i}{j-i}.$$

7.10.3   SYSTEM SETUP AND KEY GENERATION

By choosing a bilinear map, the TA first configures the SC-CP-ABE system: $f : M_0 \times M_0 \to M_1$ order of prime αp with the generator. Formerly, two arbitrary $\beta, \gamma \in Zp$ are selected by TA. The public key criteria are stated as follows:

$$\text{Public key UK} = \left(M_0, m, e = m^{\alpha}, h = m^{\frac{1}{\gamma}}, f(m,m)^{\beta} \right). \tag{7.1}$$

$\text{MK} = \left(\gamma, m^{\beta} \right)$, which the TA only identifies, is the master key.

Each user is forced to prepare with the TA, and verifies the user's characteristics and creating the user's proper private keys. An element could be of any specific string which the user assigned to it defines, classifies, or annotates. The critical discovery algorithm grosses as it inputs a group of Q attributes assigned to the user and outputs a set of essential private aspects corresponding to each of the Q attributes. The following operations are carried out by the key generation algorithm:

1. Select a random $u \in \mathbb{Z}_p$.
2. Select a random $u_i \in \mathbb{Z}_p$ For each attribute $i \in Q$.
3. To calculate the private key as:

$$\text{PK} = \left(D = m^{\frac{(\beta+u)}{\gamma}} \; ; \forall i \in Q : D_i = m^u \times E(i)^{u_i} ; D_i' = m^{u_i} \right)$$

4. PK would be sent over a protected channel to the DO.

7.10.4   ENCRYPTION

A DO must define a policy tree $T = T_{ESP} \wedge T_{Do}$, where Δ is an AND logical operator linking two sub-trees T_{ESP} and T_{Do}, to compute the encryption and protect data

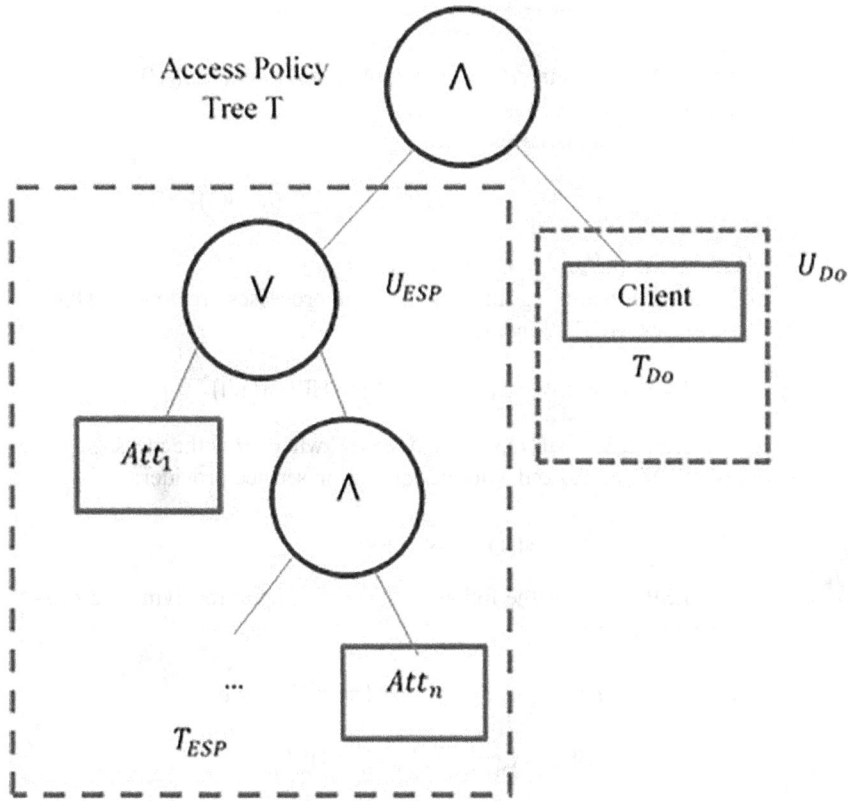

FIGURE 7.4 Access policy tree.

privacy. The T_{ESP} is a DO-controlled access policy of the data which will be enforced by the encryption service provider and T_{Do}. The privacy-preserving data and encryption computation, a DO use to specify a policy tree needs. To decrease the computing overhead at the DO, T_{Do} typically has a minimal number of attributes where specifically it should have one attribute sub-tree, as seen in Figure 7.4.

DO will arbitrarily define a one-degree polynomial $nU(a)$ and set $q = nU(0)$, $q_1 = nU(1)$, and $q_2 = nU(2)$, if T_{Do} has one attribute. DO will then give $\{q_1, T_{ESP}\}$ to ESP, which is labeled as:

$$DO \xrightarrow{\{q_1, T_{ESP}\}} ESP.$$

Here, we must remember that there is no secret of our solution will be revealed by submitting q, q_1, and T_{ESP}.

ESP then runs the Encrypt (q_1, T_{ESP}) algorithm that is designated below:
   1. $\forall a \in T_{ESP}$ arbitrarily select a polynomial n_a of degree $dg_a = ky_a - 1$, where ky_a is the hidden distribution limit significance.

a) The root of T_{ESP}, for instance U_{ESP}, selects a $dg_{U_{ESP}}$-degree polynomial with $nU_{ESP}(0) = q_1$.

b) $\forall a \in T_{ESP} \setminus U_{ESP}$ sets d_a-degree polynomial with $n_a(0) = n_{parent(a)}$ $(index(x))$.

2. To create a temporal ciphertext (XT),

$$XT_{ESP} = \{\forall w \in W_{ESP} : X_w = m^{n_w(0)}, X'_w = HF(att(w))^{n_w(0)}],$$

where W_{ESP} is the group of leaf nodes in T_{ESP}.

Between the intervening times, subsequent processes are done by DO:

1. Encrypt (q_2, T_{DO}) and derive:

$$XT_{DO} = \{\forall w \in W_2 : X_w = m^{n_w(0)}, X'_w = HF(att(w))^{n_w(0)}].$$

2. Calculate $\hat{X} = Mf(m,m)^{\beta q}$ and $X = e^q$, where M is the message.

3. Send XT_{DO}, $\hat{X}$, and X to the encryption service provider:

$$DO \xrightarrow{\{XT_{DO}, \hat{X}, X\}} ESP.$$

ESP produces the following ciphertext upon receiving the message from the DO:

$$XT = T = T_{ESP} \wedge T_{DO}; \hat{X} = Mf(m,m)^{\beta q}; X = e^q;$$

$$\forall w \in W_{ESP} \cup W_{DO} : X_w = m^{n_w(0)}; X'_w = HF(att(w))^{n_w(0)}.$$

Lastly, the service provider of encryption sends XT to the service provider of storage.

7.10.5 Decryption

Meanwhile, bilinear pairing is a comfortable undertaking. The CP-ABE decryption algorithm is computationally expensive. By farming out the costly pairing tasks to the decryption service provider, SC-CP-ABE solves this computing problem. The outsourcing would not, again, reveal the ciphertext's data material to the DSP.

The DO first conceals its private key by selecting an arbitrary $v \in Zp$ to secure the data information and then calculates $\hat{D}^v = m^{v(\beta+u)/\gamma}$. The blinded private key is denoted as PK:

$$\widehat{PK} = \left(\hat{D} = m^{v(\beta+u)/\gamma}, \forall i \in Q : D_i = m^u \cdot HF(i)^{u_i}, D'_i = m^{u_i}\right) \qquad (7.2)$$

By implementing the DSP, the DO first checks whether the access policy T can fulfill its attributes. The DO sends $\{\widehat{PK}\}$ to the DSP and asks the SSP to give the ciphertext to the decryption service provider. The storage service provider sends $XT' = \{T; X = e^q; \forall w \in W_1 \cup W_2 : X_w = m^{n_w(0)}, X'_w = HF(att(w))^{n_w(0)}\}$ and $XT' \subset XT$ to the DSP when the request is received:

$$\text{SSP} \overset{\{\text{XT}'\}}{\to} \text{DSP} \tag{7.3}$$

Once both $\left\{\widehat{PK}\right\}$ and XT' have been obtained by the DSP, it then turns the Decrypt $(\widehat{PK}, XT')$ algorithm as observed below:

1. $\forall w \in W = W_{\text{ESP}} \cup W_{\text{DO}}$, the DSP runs a repetitive DecryptNode $(XT', \widehat{PK}, U)$ function, where the root of T is U. The repetitive function is just the same as that defined in Ref. [3], and the following is done by DecryptNode $\left(XT', \widehat{PK}, U\right)$

$$\text{Decrypt node}\left(XT', \widehat{PK}, U\right) = \frac{f\left(D_j, X_w\right)}{f\left(D'_j, X'_w\right)}$$

$$= \frac{f\left(m^u \cdot \text{HF}(j)^{u_i}, m^{n_w(0)}\right)}{f(m^{u_j}, \text{HF}(j)^{n_w(0)})}$$

$$= f(m,m)^{u n_w(0)}$$

$$= H_w$$

The procedure is handled as trails: $\forall w$ is the derived node of a, and it calls DecryptNode $(XT'; \widehat{PK}; w)$ and provides the output as H_w. Let Q_a be a random ky_a-sized set of derived nodes w, and the decryption service provider figures:

$$H_a = \prod_{w \in Q_a} H_w^{\Delta_{j,Q'_a}(0)}$$

$$= \prod_{w \in Q_a} \left(f(m;m)^{u \cdot n_w(0)}\right)^{\Delta_{j,Q'_a}(0)}$$

$$= \prod_{w \in Q_a} \left(f(m;m)^{u \cdot n_{\text{parent}(w)}(\text{index}(w))}\right)^{\Delta_{j,Q'_a}(0)}$$

$$= \prod_{w \in Q_a} f(m;m)^{u \cdot n_a(j)\Delta_{jQ'_a}(0)}$$

$$= f(m,m)^{u n_a(0)}, \tag{7.4}$$

Here $j = \text{index}(z)$ and $Q'_a = \left\{\text{index}(z) : z \in Q_a\right\}$.

Finally, the recursive algorithm returns $O = f(m,m)^{uq}$.

2. Compute

$$f\left(X, \hat{D}\right) = f\left(e^q, m^{\frac{v(\beta+u)}{\gamma}}\right) = f(m,m)^{vuq} \cdot f(m,m)^{v\beta q}$$

3. Send $\left\{ O = f(m,m)^{uq}, L = f\left(X,\hat{D}\right) = f(m,m)^{vuq} \cdot f(m,m)^{v\beta q} \right\}$ to the DO:

$$\text{DSP} \overset{\{O,L\}}{\rightarrow} \text{DO}.$$

On receiving $\{O,L\}$, DO calculates $L' = L^{\frac{1}{v}} = (m,m)^{uq} \cdot f(m,m)^{\beta q}$ and then it recovers the message:

$$M = \frac{\hat{X}}{(L'/O)} = \frac{Mf(m,m)^{\beta q}}{\left(f(m,m)^{uq} \cdot f(m,m)^{\beta q} \right) / f(m,m)^{uq}}$$

7.11 ATTRIBUTE-BASED DATA RETRIEVAL

SC-CP-ABE-based ABDR scheme is used for adequate, accessible data controlling and allocation.

7.11.1 DATA MANAGEMENT OVERVIEW

Repeated data upgrades can result in extra costs for maintaining files. For example, to enhance the existing files, modify some field of data in encoded data where the encoded data need to be transmitted for decryption from storage service provider to decryption service provider. After the updates are done, it is essential to re-encrypt the ESP and transmit it to the storage service provider. Therefore, the re-encoding process needs the data to be retrieved and uploaded, which will involve high overhead interaction and processing and would cost more for DOs as a consequence.

The cost problem is resolved by breaking a file into several blocks, which are encoded dynamically. The DO will import the relevant blocks to be changed to upgrade files. We may prevent re-encrypting the entire data in this manner. In comparison, data access protection can be implemented using "lazy" re-encryption techniques on individual blocks. For example, suppose data access memberships are modified (i.e., the access tree is switched) to a specific file. In that case, this occurrence may be registered, but no modifications to the file are initiated. The suggested SC-CP-ABE system then achieves the re-encoding before the content of data has to be changed.

Addition overhead is often applied by separating the data into several small chunks. It's because for every code block for data controlling, additional controller information has to be connected. For starters, a chunk ID and an indicator to its respective access of data T (tree) should be included in the control message. We show an example file contained in the storage service provider in Figure 7.5. Any file is segmented, as seen in Figure 7.4. A chunk is a {Block ID, Pt, Data which are encrypted} tuple, where Block ID is the particular block identifier; Pt is the XT control block pointer; and the encrypted data are with a KED. A {CID, Encrypted KED} block of control has an ID of the CB (control block), i.e., IDC and KED encrypted with the SC-CP-ABE scheme.

The difficulty of the data block's acceptable size to be separated from an identified file size is in the ABDR method. In this trade, the aim is to minimize the overhead area and coordination with the resulting basic expectations in mind:

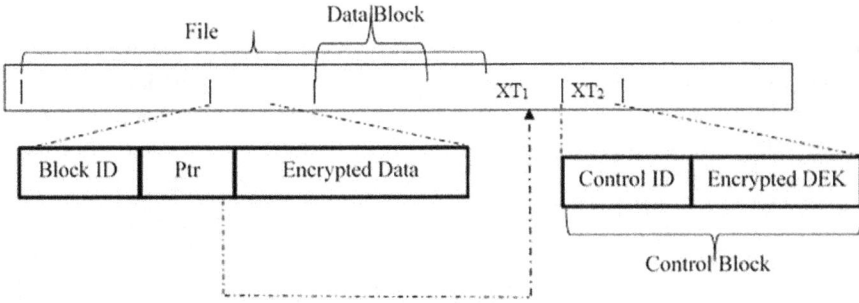

FIGURE 7.5 Sample file storage model.

1. And data change can affect only a limited amount of data, e.g., the upgrading of some database data fields.
2. The block of data change is known in each unit field.
3. Each data block will have the same possibility of updating.

Collecting traffic information, where lightweight instruments such as cell phones and devices act as movable or static data collection proxies, is an outstanding implementation scenario that follows those assumptions. The systems change the corresponding data fields in the encrypted databases regularly.

The overall cost TC in the meantime is as follows:

$$TC = 2n_b D_c + \frac{F_s}{B_s} C_s S_c \tag{7.5}$$

where n_b is the amount of revised chunks in a span of meantime and $2n_b$ means to upgrade entails single encoding that takes two broadcasts; B_s is the block size; D_c is the data transmission cost rate that is indicated by virtual storage and cellular service providers; F_s is the file size; C_s is the controller data size for every data chunk; and S_c is the storage accusing rate. To minimize the total cost TC, DO will reduce (5) and develop the best chunk size:

$$B_s \geq 2\sqrt{2n_b D_c F_s C_s S_c}$$

7.12 PERFORMANCE EVALUATION

7.12.1 COMPUTATION PERFORMANCE OF SC-CP-ABE

To determine the efficiency of the presented SC-CP-ABE process, service operators and consumers' computational costs are focused on in both conceptual analysis and empirical results.

First, we examine the amount of costly cryptographic operations performed by service providers and computers of users over M_0 and M_1, i.e., matching, exponentiation, and multiplication. We presume that the access policy T_{ESP} has a_1 attributes linked by an AND logical gate in our study and T_{DO} has only one attribute. Furthermore, an AND gate is the root node. We measured the number of exponentiations, increases, and hashes sustained on the encryption service provider side and client side in the encryption servicing in the following table with M_0 operations, where x_1 is the number of attributes in T_{ESP}:

	Exponential M_0/M_1	Multiplication M_1	Hashing to M_1
ESP	$2x_1/0$	0	x_1
Data user	$2/1$	1	1

Shown in the following table is a measure of the number of exponentiations, increases, inversion, and pairing operations incurred through decryption servicing on the decryption service provider and user's hand, where x_1 is the number of T_{DSP} attributes:

	Exponential M_0/M_1	Multiplication M_1	Inverse M_1	Pairing
DSP	x_1	$2x_1$	x_1	$2x_1 + 1$
Data user	1	2	1	0

Here, mobile devices, TPA, and the cloud service provider execute associated functions. A mobile node does more encoding than decoding, while TPA accomplishes both encryption process functions, decreasing the mobile handler's functionality. No CSP encryption/decryption is carried out here. The CSP also stores the protected file. Table 7.2 demonstrates the overhead calculation of the cell terminal, TPA, and cloud storage provider. First, the quantity of inexpensive cryptographic operations is measured, i.e., incremental functions, grouping functions, and the number of hash functions. The handheld terminal and TPA are used to execute these cryptographic operations. An association of incremental functions, hash functions, and grouping functions scheduled on mobile devices and TPA during encryption and decryption is seen here in the table.

TABLE 7.2
Computation Overhead of TPA

	Exponential M_0/M_1	Hashing to M_1	Pairing
Encryption Process			
TPA	1	0	1
Data user	3	1	0
Decryption Process			
TPA	2	1	0
Data user	1	0	0

Next, the exponential operations conducted during the encryption process on the mobile terminal are evaluated, and then on TPA. First, the incremental function is accomplished during the encoding process of the data. After that, a single incremental function is applied while the encoded file's re-encoding process and one incremental function are implemented during hash encoding. Thus, an aggregate of three incremental functions are accomplished on the cellular device. Currently, during the decoding process of the received message, one incremental function is implemented on TPA and another incremental function is executed before storing it on the mobile cloud during the consequent decoding of the message. Thus, in encryption, TPA performs a total of two incremental functions. Just one incremental function is conducted on the cell terminal during the decryption process to decode the document. An amount of two incremental functions are performed while on TPA. This indicates that around 99% of the effort is done on TPA during decryption.

Second, it analyzes the complete hash functions done on both the cellular node and TPA. To figure out the post's hash, only one hash function is done on the user side. No hash function during the encoding function is conducted on TPA. No hash function is applied on the user side during decryption, although the hash function is done on TPA to carry out reliability checking. Third, the pairing operations carried out on the user side and the TPA side are evaluated. Here, only one grouping function is applied during encryption and one grouping function is applied at TPA. At the same time, no grouping function is done during decryption on the user side and TPA.

It can be examined from this table that no operation is conducted by cloud storage service on cloud; CSP only offers the mobile customer with the data storage facility. This eliminates the work of smartphone users, and many activities are outsourced to the TPA. The experimental validation test findings are shown in Figure 7.6, where user-side encryption and decryption computation overheads are contrasted with server-side encoding and decoding. The overhead equation is measured in log 12, i.e., based on thousands of ticks of (K) clocks. As shown in Figure 7.6, more than 89% of encryption and more than 98% of decryption are done by the access providers.

7.12.2 Storage Performance of ABDR

To evaluate and equate ABDR storage efficiency to many similar cryptography methods, transmission encoding schemes (subset difference), BGW transmission encoding, and polynomial control scheme (PCS).

The outcome is analyzed in the context of encrypted storage payload and key processing payload (input variables and cryptographic keys and TPA stored by the client). The overall amount of clients with N_u system is designated, and a client needs to share a file with any specified group of recipients in the scheme. The relative outcomes are attained in Table 7.3.

7.13 CIPHERTEXT STORAGE OVERHEAD

The ciphertext size in the subset difference scheme is $O\left(h^2 \cdot \log^2 h \cdot \log N_u\right)$, with the cumulative number of users planning to support the ciphertext. The ciphertext

Encryption/Decryption at CSP

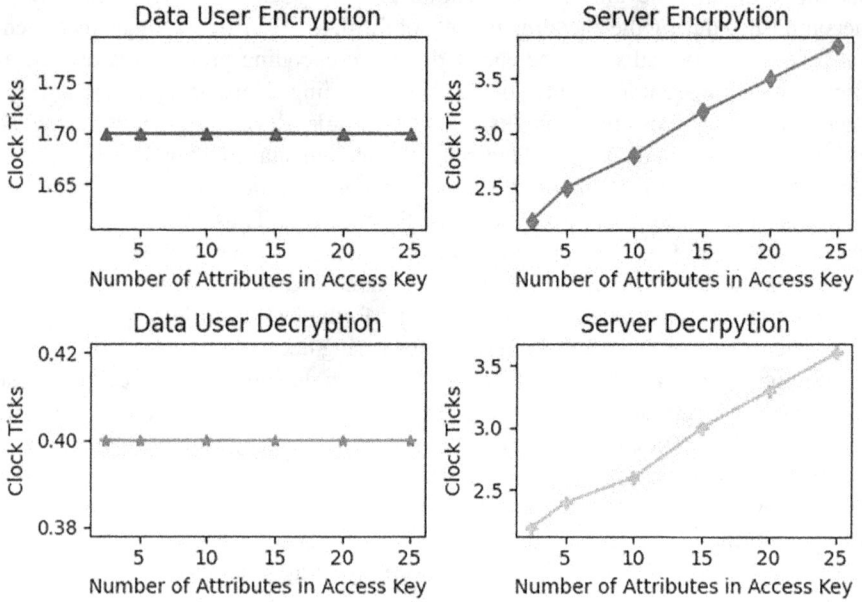

FIGURE 7.6 Comparison of user computation overhead with server.

TABLE 7.3

Ciphertext Storage Overhead and Key Storage Overhead Comparison

Scheme	Ciphertext Storage		Key Storage	
	One Data Receiver	Multiple Data Receivers	TPA	Data User
ABDS	$O(\log N_u)$	$O\left(\log^2 N_u\right)$	$O(1)$	$O(\log N_u)$
Subset difference	$O\left(h^2 \cdot \log^2 h \cdot \log N_u\right)$	$O\left(h^2 \cdot \log^2 h \cdot \log N_u\right)$	$O(N_u)$	$O\left(h \cdot \log h \cdot \log \cdot N_u\right)$
BGW	$O(1)$	$O(1)$	NA	$O(N_u)$
ACP	$O(N_u)$	$O(N_u)$	$O(N_u)$	$O(1)$

---

size for the BGW system is $O(1)$ or $O\left(N_u^{\frac{1}{2}}\right)$. In the PCS system, the message size depends on the amount of polynomial control scheme, which is proportional to recent recipients' number. Thus, $O(N_u)$ is referred to as the message scale. The ciphertext size depends upon the number of commodity terms in $S_Q^{\min}$ to manage a collection of receivers using ABDR. In a minimized sum of product, the writers obtained a more excellent bound and an inferior inevitable on the typical product terms. In experimental words, $\approx \log(N_u)$ is the average number of necessary texts.

The experimental results of ABDR in a system with 512 and 1024 users to examine the typical scenario and the amount of information needed are shown in Figures 7.7 and 7.8,respectively. In the experiment, we deliberate cases not allocated to 0%, 20%, 30%, and 50% ID. Different numbers of receivers are selected randomly from the category for each event. We repeat the findings 100 times to average them. The message size in CP-ABE begins experimentally at around 680 bytes, and each extra attribute includes around 320 bytes. Since the amount of features is restricted by log N_u in the access policy, we can infer that the expense of ABDS ciphertext storage is in the order of $O(\log 2N_u)$.

7.14 KEY STORAGE OVERHEAD

ABDR significantly lowered the key management overhead of consumers' TPA and computers relative to transmitted encryption systems. In ABDR, the scales of the SK and MK are constant. Often, assignment attributes of $\log(N_u)$ bits need to be saved by a user. The storage overhead is, therefore, $O(\log N_u)$, assuming that a customer stores no IDs of the data receivers. Since the DO may be essential the incline of data recipient IDs beside the collection of IDs that do not care about prioritization of Boolean features, we can claim that it does not require additional overhead storage.

- After the transmission, data publishers do not have to hold the receiver's IDs; the data storage will then be published.
- The TA can be an ongoing process which reduced the Conventional of all non-care IDs that can further decrease the number of messages by data publishers.
- If consumers are successively allocated IDs, i.e., through low to high, TA will issue the minor unassigned IDs to all users, who can use the highest IDs as values do not matter.
- If the consumer desires to store the IDs, the storage space is simply $N_u \log N_u$ bits if $N_u = 2^{18}$.
- When a data publisher does not care ideals, which decrease the association attribute type, the above ciphertext storage may be marginally higher.

As shown in Figures 7.7 and 7.8, the 0% vacancy curve may be seen as the overhead ciphertext storage needed if IDs are not identified or cared for by data publishers.

Mobile terminal and TPA: Necessary storage requirements for mobile terminals and TPA is as follows. The mobile terminal only holds the static size key here, so the complexity of keeping this key is constant, which is $O(1)$. Although TPA stores the key and the hash of a file, they are greater than the storage complexity for mobile stations.

Participants	Storage requirements
Mobile device	Contains public key of TPA
TPA	Contains key(public/private) of TPA and a hash value of the function conventional from the user
CSP	A ciphered file of the user

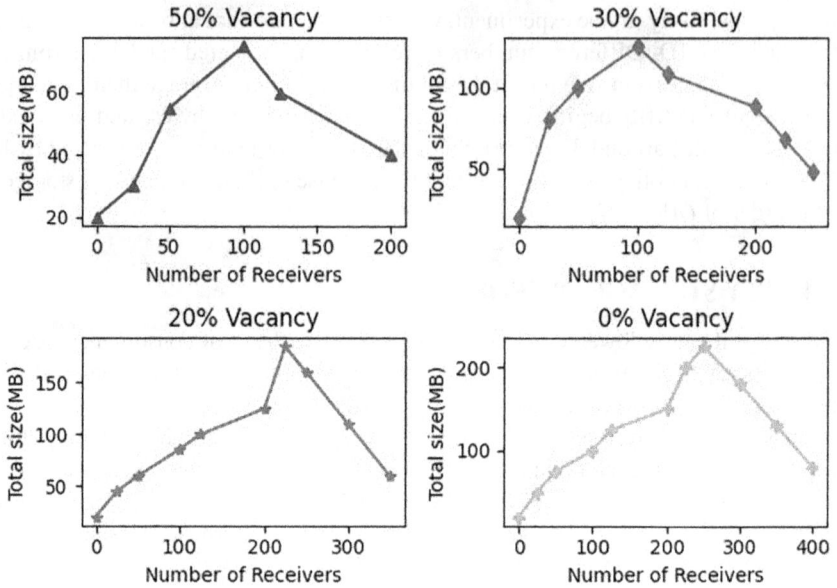

FIGURE 7.7   Ciphertext size for group size 512.

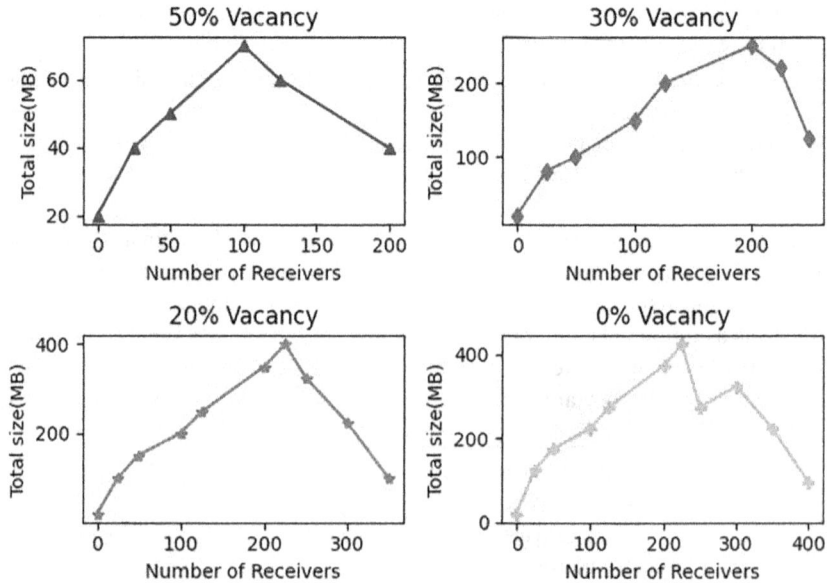

FIGURE 7.8   Ciphertext size for group size 1,024.

Here, the storage of cloud service has many storage capacities for consumer file storage; then, there is no need to think and stress about CSP storage requirements. The mobile system only needs keys to be kept, but it is not a huge agreement for mobile devices since keys are steady in size. As mentioned above, decryption procedures are offloaded onto TPA. The strenuous effort of reliability checking and encoding removes many of the smartphone user's work and makes this suggestion more successful. The encrypted data are hardly kept on the cloud servers here, so the suggested framework understood storage capacity and processing power's objective.

7.15 CONCLUSIONS

Suppose the information is stored on the cloud by a mobile device restricted by a resource. There is also a big issue about the cloud service providers storing the documents properly. In mobile cloud storage, stability is the principal concern. The suggested structure offers a reliable architecture for cloud data storage services in demand to protect data processing in mobile-based cloud computing. Specifically, the proposed result allows portable wireless gadgets to supply and recover their data safely and at a reduced cost in MCC. This chapter suggested a new SC-CP-ABE (secrecy-conserving ciphertext-policy attribute-based encryption) to secure encoded data from users. Lightweight computers can quickly offload specific encoding and decoding practices to the provider of cloud service using SC-CP-ABE without exposing the data content and security keys used. As a cryptographic access management method, we proposed an attribute-based data retrieval (ABDR) framework. Theoretically, ABDR achieves optimum data to reduce overhead costs for calculation, storage space, and connectivity. Specifically, as well as coordination overhead for data storage, ABDR minimizes cloud costs paid by cloud service providers. Our performance evaluations indicate our solution's security intensity and usefulness regarding computing, connectivity, and storage. In reality, SC-CP-ABE relies on the BSW CP-ABE method, which is influenced by sequential growth in the amount of the ciphertext. We studied a CP-ABE method of constant ciphertext size and attempted to suggest a sourcing mechanism for the current CP-ABE scheme that protects privacy. Encryption is used in this system to provide the data with protection during transmission. Since the encrypted file is saved in the cloud, users may presume that their information is safe. The scheme register is transmitted over the channel only in encrypted form, which eliminates the issue of leakage of information. No, the file may be accessed by a third party or attacker since that entity does not recognize the DO's key. In any area of work, there is still a potential for improvement, so here too. These are some of the predictions made in all protection models that TPA is unbiased. All observations and reviews are forwarded to TPA, so to allow TPA to be more secure, there is a necessity to do some research. Other frameworks used in the mobile cloud storage environment for secure storage providers may be explored in future research. Such work can also be done to decrease the workload of intelligent devices. In turn, advance performance enhancements can be accomplished by precompiling and secreting some commonly used access policy trees from encryption service provider. The additional significant upcoming task will be introducing a reliable consumer storage file system based on common public cloud storage so that consumers can easily protect their cloud storage.

REFERENCES

1. Zhou, Z., D. Huang, and Z. Wang. "Efficient privacy-preserving ciphertext-policy attribute based-encryption and broadcast encryption." *IEEE Transactions on Computers* 64, no. 1 (2013): 126–138.
2. Zhou, Z. and D. Huang. "Efficient and secure data storage operations for mobile cloud computing." *In 2012 8th International Conference on Network and Service Management (CNSM) and 2012 Workshop on Systems Virtualization Management (SVM)*, pp. 37–45, IEEE (2012), (22 -26 October 2012, held at The Mirage (Las Vegas), USA.
3. Garg, P. and V. Sharma. "An efficient and secure data storage in mobile cloud computing through RSA and Hash function." *In 2014 International Conference on Issues and Challenges in Intelligent Computing Techniques (ICICT)*, pp. 334–339, IEEE, Held 7-8 February 2014, Ghaziabad, India.
4. Pranav, P. and N. Rizvi. "Security in mobile cloud computing: a review." *International Journal of Computer Science and Information Technologies* 7, no. 1 (2016): 34–39.
5. Matilda, S. and T. Ananth Kumar. "12 the winning combo." In E. G. Julie, J. J. V. Nayahi, N. Z. Jhanjhi (Eds) *Blockchain Technology: Fundamentals, Applications, and Case Studies*. Boca Raton, FL: CRC Press (2020): 199.
6. De, D. *Mobile Cloud Computing: Architectures, Algorithms and Applications*. Boca Raton, FL: CRC Press (2016).
7. Xu, J., H. Hong, G. Lin, and Z. Sun. "A new inter-domain information sharing smart system based on ABSES in SDN." *IEEE Access* 6 (2018): 12790–12799.
8. Kumar, K. S., T. Ananth Kumar, A. S. Radhamani, and S. Sundaresan. "Blockchain technology: an insight into architecture, use cases, and its application with industrial IoT and big data." In E. Golden Julie, J. Jesu Vedha Nayahi, Noor Zaman Jhanjhi (Eds) *Blockchain Technology*. Boca Raton, FL: CRC Press (2020): 23–42.
9. Guan, Z., T. Yang, and X. Du. "Achieving secure and efficient data access control for cloud-integrated body sensor networks." *International Journal of Distributed Sensor Networks* 11, no. 8 (2015): 101287.
10. Ramu, G. and B. Eswara Reddy. "Secure architecture to manage EHR's in the cloud using SSE and ABE." *Health and Technology* 5, no. 3–4 (2015): 195–205.
11. Adimoolam, M., A. John, N. M. Balamurugan, and T. Ananth Kumar. "Green ICT communication, networking and data processing." *In Green Computing in Smart Cities: Simulation and Techniques*, pp. 95–124, Springer, Cham, (2021).
12. Saikeerthana, R. and A. Umamakeswari. "Secure data storage and data retrieval in cloud storage using cipher policy attribute-based encryption." *Indian Journal of Science and Technology* 8, no. S9 (2015): 318–325.
13. Ibraimi, L., M. Petkovic, S. Nikova, P. Hartel, and W. Jonker. "Mediated ciphertext-policy attribute-based encryption and its application." *In International Workshop on Information Security Applications*, pp. 309–323, Springer, Berlin, Heidelberg (2009).
14. Usharani, S. and R. Ramya. "Security-based novel context-aware mobile computing scheme via crowdsourcing." (2017): 2395–1990.
15. Bala, P. M., S. Usharani and M. Aswin. "IDS based fake content detection on social network using bloom filtering." *In 2020 International Conference on System, Computation, Automation and Networking (ICSCAN)*, pp. 1–6, Pondicherry, India (2020). doi: 10.1109/ICSCAN49426.2020.9262360.
16. Li, J., Q. Wang, C. Wang, and K. Ren. "Enhancing attribute-based encryption with attribute hierarchy." *Mobile Networks and Applications* 16, no. 5 (2011): 553–561.

8 Deep Dive on Popular Security Models and Strategies of Cloud Computing

K. Venkatesh
Vel Tech Rangarajan Dr. Sagunthala R&D
Institute of Science and Technology

R. Velvizhi, K. Selva Banu Priya, and N. Sivaranjani
Bharath Institute of Higher Education and Research

CONTENTS

DOI: 10.1201/9781003219880-8

8.1   ACCESS CONTROL MECHANISMS IN CLOUD COMPUTING

Cloud computing is a gradually developing technology in the field of information technology (IT). It is Internet-focused computing [1] which provides numerous reliable services to users to preserve privacy, security, confidentiality, integrity, authentication, etc. It also keeps safeguarding from various vulnerable and malicious attacks [2]. Utilizing the access control principles, the access control mechanism takes numerous attributes. Access control mechanism usually has user's features, which are fundamental functions used to perform various appropriate procedures, functionalities, and suitable characteristic functions rather than the procedures performed in isolation in multiple fields. Hence, the greater importance of the system had been evaluated quickly [3,4]. Consequently, employing activating or contradicting the preferences being assessed based on the set of rules. The access control mechanism follows this set of rules. While finding the likes of access control, five various entities have to be considered [5,6].

1. It allows appropriate users to protect the cloud ecosystem.
2. It accesses the node's or system's confidentiality.
3. It is functioning in the process of monitoring the network privacy sequence.
4. Also, it ensures that the sequence of functions is processed correspondingly.
5. The need to implement a control system in a static series also incorporates preferences.

Access control mechanism provides authentication to the end-users to authorize the resources are available publicly to them. In recent decades, numerous access control mechanisms have been made to secure data access. This mechanism provides security to the node and enables access to the connected object to the network. Traditional access control mechanism comprises three main categories that include discretionary access control mechanism (DACM), mandatory access control mechanism (MACM), and role-based access control mechanism (RBACM). The primary purpose of inventing access control mechanisms in the cloud platform is to protect unauthorized users from the network and enable the network to be much stronger in the privacy aspect.

Traditional access control mechanism enables authentication database and examines the reference monitor. In the process of automation and monitoring, it is widely utilized. The privacy of nodes in a network influences the organization either directly or indirectly. It is also said to be the procedure or policy that permits, declines, or hinders authorization to the node in a network and is used to point out identity-based access control mechanism [7]. All the storage-related functionality is done through cloud storage and emphasizes the privacy control action and access options. It is not a way out for data security. With the help of data, encryption is enabled to maintain data security.

8.1.1 DISCRETIONARY ACCESS CONTROL MECHANISM

It is one of the traditional access control paradigms, in which the overall control of the program is with the user. DACM is termed as providing authorization to the node in a network which also enables identity to operate on available policies. Moreover, it is used to evaluate permission to the user to a corresponding object. DACM is mainly based on user and authentication identity that particularizes for every node that accesses object and method that the user pleads (Figure 8.1).

In this mechanism, the authentication is expressed explicitly, and also separate user authorization has been closed. Suppose authentication is open, it is known as open stratagem. DACM comprises both access attributes and rules. Access attributes permit the nodes to describe the various authentication functionalities and enable the prevention safety mechanism that reduces unauthorized users from the cloud environment. Furthermore, permission is granted by the owner of an object, and each user seeks permission from that owner. Each object in a file system is associated with an access control list. Simple discretionary access control has the following four features: the access control mechanism's fundamental actions.

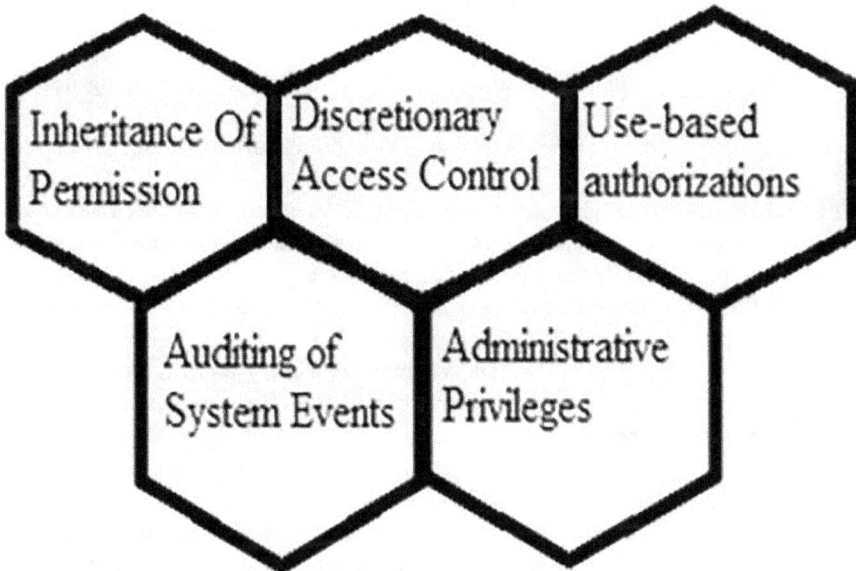

FIGURE 8.1   Discretionary access control.

a. Inheritance of permissions
b. User-based authorizations
c. Auditing of systems
d. Administrative privileges.

Pros of DACM

1. This mechanism was functioning on the authentication database principle, which includes numerous users that maintain numbers of records.
2. DACM also enhances the malleability of usage of the information.

Cons of DACM

1. In DACM, there is no limitation for the information that has been accessed. Also, no guarantee for the information that has been ruined around the cloud environment. This creates uncertainty on the accessibility of information and also has a chance for information loss.
2. There is a chance to enter malicious third parties in a node that makes the network vulnerable; also, information is not accessed based on consistency.
3. This makes the chance to steal all the original information without the knowledge of owners.

8.1.2 MANDATORY ACCESS CONTROL MECHANISM

Mandatory access control mechanism (MACM) is focused on the approach that consists of the number of subjects corresponding to the objects. MACM is mainly dependent upon the privacy level of the network. Traditional MACM is primarily for the security parameter to achieve some privacy control mechanism. This maintains two important principles [8].

1. User's privacy function dominates access of object during reading (read down).
2. User's privacy function dominates entrance of object during writing (write up).

MACM depends on the object classification and a related subject in the cloud ecosystem. Every object present in the cloud environment must satisfy some fundamental relationship, and then only the access is granted. Correspondingly, everything and its subject in the cloud ecosystem adopt some privacy function. This privacy function utilizes to find out the object's current access state.

In such a way, the privacy function accompanying every node in the cloud environment is clearance [8]. In MACM, the node in a network cannot modify the access control and privacy function. Without suitable authentication, MACM will not allow users within the network and enhance high-level security functions to provide an ecosystem from various malicious and vulnerable attacks.

Figure 8.2 summarizes that the mandatory access control mechanism is determined based on privacy function. The privacy function hierarchically depends on the privacy control mechanism and arranged from the highest privacy level to the lowest privacy level. In this instance, Alice and Bob are doing a privacy function in which MACM

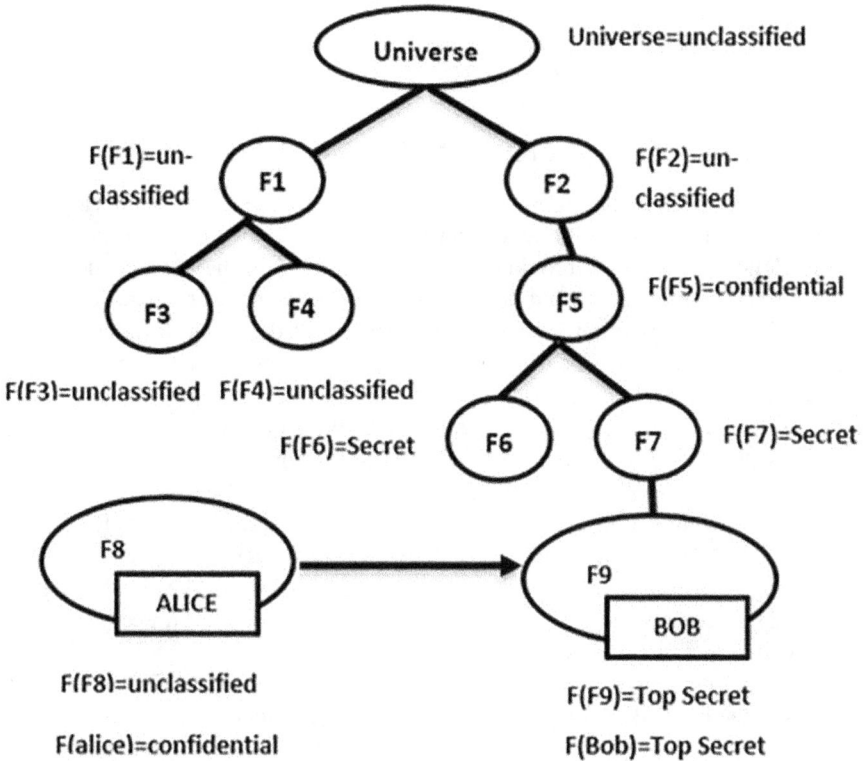

FIGURE 8.2   Mandatory access control mechanism.

classification is maintained. The classification hierarchy follows a systematic approach to maintain information confidentiality and integrity, i.e. Top Secret (TS), Secret (S), Confidential (C), and Unclassified (U) [8]. Alice has the authorization to access secret information from the cloud ecosystem, in which confidentiality is maintained. But Alice has no rights or authorization to access or read that secret information since Bob is maintaining a TS function. In this instance, the subjects are said to be Alice and Bob. The object is said to be the precise information they request. In this organized way, an information process has been done to exchange and enhance the activities.

Pros of MACM:
1. It enhances high-level integrity to prevent the flow of information from low-level to high-level objects. Integrity achieves this information sequence.
2. MACM provides more comprehensive applications in the field of military and hospitality.
3. It also enhances privacy function in a multilevel.
4. In MACM, each user's response will be intermediate, so the cloud environment is more secure to access the information. When compared to other access control mechanisms, MACM achieves lower scalability.

Cons of MAC:

It does not modify the privacy function so that once we identify a particular subject, we can't catch it.

8.1.3 ROLE-BASED ACCESS CONTROL MECHANISM

Employing takes the decision; role-based access control mechanism (RBACM) considers an individual's character and trustworthiness of every action involved in the cloud ecosystem. It provides users' accessibility to the network depending on the user's activity in a cloud ecosystem. To identify the user's role, RBACM is needed. The user's position is based on the priority of the network function. Roles and responsibilities are allocated on the cloud ecosystem schema along with various security measures. Every role in the cloud ecosystem depends on the authenticated user, systematic commands, and dedicated information access. It depends on least privilege only; roles can assign. Functions can organize by an intermediate mechanism.

RBACM is implemented using four main design constraints such as RBACM-0, RBACM-1, RBACM-2, and RBACM-3. Simultaneously, RBACM permits users to perform numerous activities, and those activities help determine the cloud ecosystem effectively. In some scenarios, one activity is allocated to one user, and it is used to identify the exact function done by other users (Table 8.1).

Figure 8.3 demonstrates the overall performance of RBACM. Every user who belongs to the network, say node-1, node-2, node-3, ..., node-n, must have a unique role function. Roles are like an activity. They are done in the entire mechanism. Every operation is grant permitted by the system administrator to a particular user. Each process must have some object; through that object, only it accesses the sequence of functions. RBACM provides the exact mechanism of securing the data from the web to give the functionalities in the cloud ecosystem. It also utilizes and assigns various roles to each user in the environment. Through user-pull architecture and server-pull architecture, RBACM is implemented [10].

1. **User-pull architecture:** Every node pulls its function from the environment.
2. **Server-pull architecture:** Every node associated with the web server in such a way removes their function coordinated to the environment.

TABLE 8.1

Various Role-Based Access Control Mechanisms and Their Role

Control Mechanism	Role
RBACM-0	It depends on the privileges at least and its separation. It cannot hold any access, and hierarchy to the corresponding object is allocated directly [9]
RBACM-1	It depends on hierarchies and the principle used for use cases
RBACM-2	It is based on the hierarchy along RBACM-1
RBACM-3	It is based on both hierarchies along with corresponding constraints

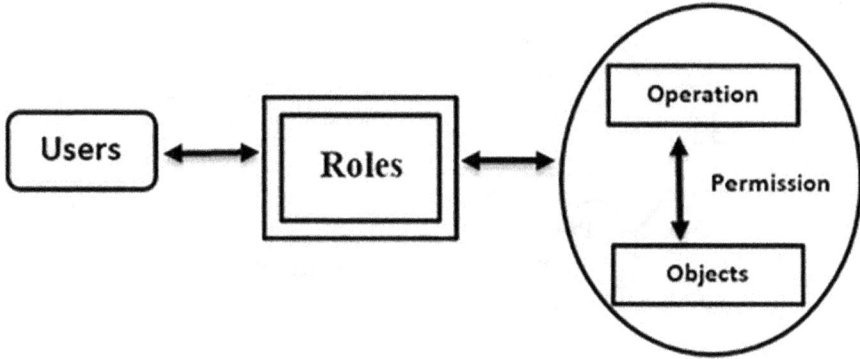

FIGURE 8.3   Role-based access control mechanism.

Pros of Role-Based Access Control Mechanism:
1. It provides a hierarchy-based mechanism, so everything must be arranged in a sequence.
2. Role allocation is based on least privilege, reducing the leakage or modification of information done by intruders. Based on executing cloud environment only, it allows users to depend on the classification.

Cons of Role-Based Access Control Mechanism:
i. Sometimes, identification of the user's role is challenging.
ii. The role provided by every user depends on their privilege, but there is confusion while changing the function to get permission, where the procedure is allocated, etc. (Figure 8.4).

8.2   SECURE DEPLOYMENT STRATEGIES OF CLOUD PLATFORM

A cloud deployment strategy is how the cloud provides services and reaches out to the network node. It paves the way for accessing the web with enhanced security, privacy, and integrity.

8.2.1   PUBLIC CLOUD

Public cloud grants system and services to be effortlessly maintained and authorized to provide its functionality to the general public. Because of its openness, on consideration of privacy, it is less secure, e.g. email. It is also considered to be the dominant strategy of the cloud environment. It enables all general services and functionalities over the network in an open, unsecured environment, e.g. Google Search and Amazon EC2 [14].

8.2.2   PRIVATE CLOUD

Private cloud grants system and services effortlessly within a particular organization, and third-party access cannot be granted. Several aspects have been followed

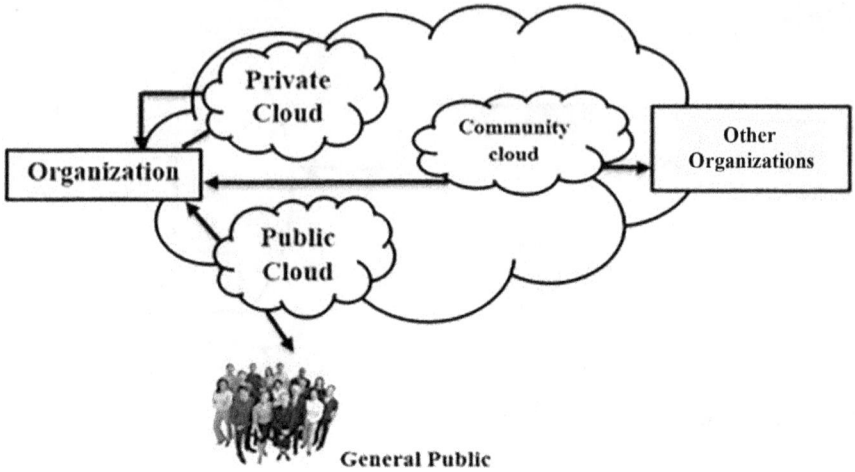

FIGURE 8.4   Deployment model for cloud environment.

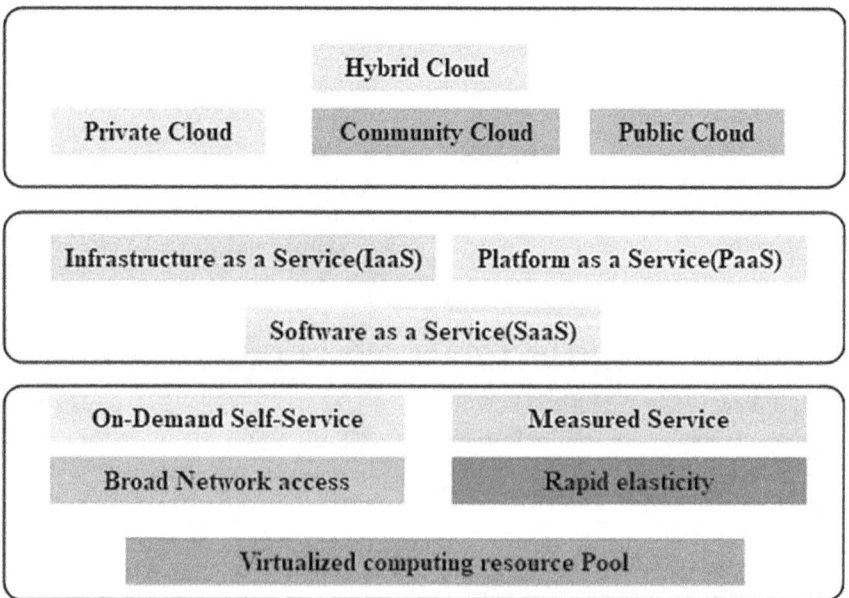

FIGURE 8.5   Deployment model based on providing services for cloud environment [18].

to provide this set-up within the organization: initially to improvise and optimize the usage of the existing cloud ecosystem environment and then to enable security, which means maintaining privacy function, various parameters have been held to achieve security wisely after comparing them with other infrastructures, whether applicable to keeping and functioning all the security concerns. Then, finally, the network is accessed behind its corresponding firewalls (Figure 8.5).

8.2.3 Community Cloud

Community cloud enables various functionalities easily within a group of networks or organizations. It has all the access to modify or involve all the functions and activities within that particular group. It could be actioned on the third-party vendor or a specific organization in the group or community.

8.2.4 Hybrid Cloud

The combination of public, private, and community cloud is known as hybrid cloud. It enables all sequences of functions to standardize and allow data from various functionalities. It also provides the way to achieve a network that satisfies all kinds of services, including infrastructure, platform, software, on-demand self-service, and measured service.

8.3 NOVEL KEY MANAGEMENT PRACTICES IN CLOUD

For maintaining privacy in the cloud environment, a cryptographic key is used. The cryptographic key is said to be the detailed information used to find out the output's function of an algorithm, e.g. symmetric key or asymmetric key. It can be used to keep the data more confidential and systematically achieve security, for instance, in a cloud environment consisting of multiple users. Each user must create a key with the help of key management system (KMS) [21]. If the key gets crashed, KMS helps to get the key back. Also, if the key is not used for any functioning, KMS automatically deletes the key. It functions based on metadata. Metadata holds all the information about crucial identifier, key label and cryptographic algorithm, key's length, and usage count of crucial and parameter functions. The life cycle of critical includes the following stages (Figure 8.6):

 i. Creation
 ii. Initialization
 iii. Distribution

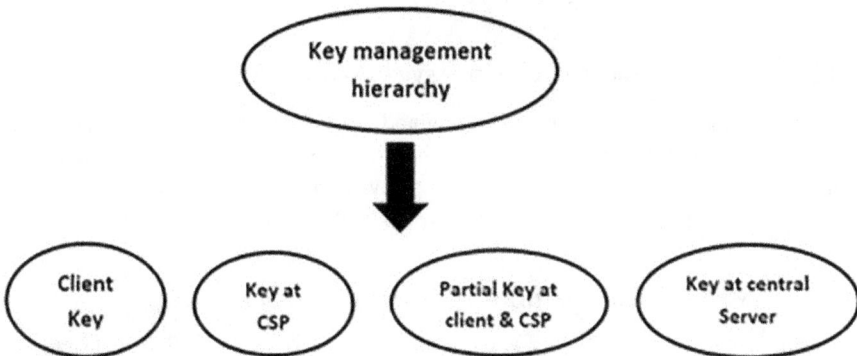

FIGURE 8.6   Key management hierarchy.

FIGURE 8.7   Key management at client side.

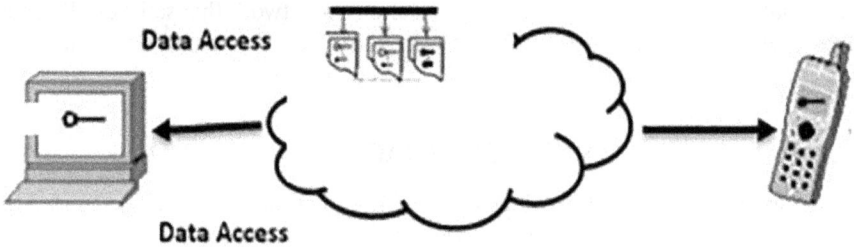

FIGURE 8.8   Key management at cloud service provider.

   iv) Active state
    v) Inactive state
  vi) Termination.

8.3.1   KEY MANAGEMENT AT CLIENT SIDE

In this technique, encryption is maintained to store and process data at the client side. The client might have thin, e.g. mobile phone. Keys will be enabled and kept at the customer's side. This technique is said to be a cryptographic homomorphism technique. Figure 8.7 shows the functionality of key management on the client side. At the server side, all the operations have been done in an encrypted form.

8.3.2   KEY MANAGEMENT AT CLOUD SERVICE PROVIDER SIDE

In this technique, keys are arranged, maintained, and monitored at cloud side, in which the cloud service provider functions its entire process. If the key gets lost, there is no chance to read the data available in the cloud environment. Usually, data are stored in an encrypted format; the data can be decrypted back only using the key (Figure 8.8).

8.3.3   KEY MANAGEMENT AT BOTH USER AND CLOUD SIDES

In this management, the technique key is divided into two sequences. The first sequence is stored at the user side, and another series is held at the cloud side. If both arrangements are together, then we can adequately retrieve the data. Thus, data can be controlled by

Data Access

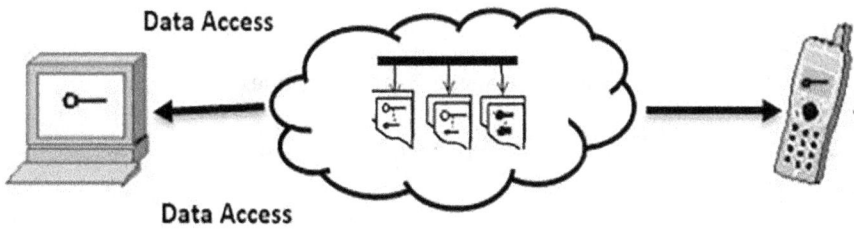

Data Access

FIGURE 8.9 Key management at both user and cloud sides.

the user and be more secure. The cloud service provider and the user do not maintain complete information about the key at the cloud side. If some information or part of that key gets lost, the actual data cannot be accessed and recovered [21] (Figure 8.9).

8.3.4 Key Splitting Technique

To access data from the cloud, the content provider will enhance and share the data. Key is distributed and split based on the users. If a particular user needs to get access to the cloud environment, then they need a key. Based on k, n key is selected. Then the user has access to encrypt and decrypt those data [21].

8.3.5 Key Management at Centralized Server

Based on the asymmetric integral approach, this technique works. By use of the public key, data get encrypted. Cloud side data are stored in the encrypted format. The user will have all permissions to access the data. The data decryption has been maintained utilizing the private key [21].

8.3.6 Group Key Management for Cloud Data Storage

In the cloud environment, with the help of trusted group members the data are shared. For securing data in a cloud environment, a group key is established. For each user, partial resolution is maintained. If a particular person needs to access the data, they can access using the group. If an individual member leaves the group, group key is generated again. If a particular member wants to join the group, they will make the key for them [21] (Table 8.2).

8.4 PRIVACY PRESERVATION OF CLOUD COMPUTING ENVIRONMENT

Cloud computing enables numerous services to users in all aspects and provides an encouraging environment for performing numerous activities, maintaining security infrastructure, and deploying various services offered by a domain. It also enhances more comprehensive applications in the IT sector to build their hardware and software independently. In such a way, a considerable amount of capital cost is saved.

TABLE 8.2

Comparison between Various Management Techniques

Various Properties and Methods	Security	Scalability	Fault Tolerance	Suitable Cryptographic Algorithm
Key management at client side	Agree	Agree	Disagree	Symmetric algorithm
Key management at cloud service provider side	Agree	Disagree	Disagree	Asymmetric algorithm
Key management at both user and cloud sides	Agree	Agree	Disagree	Symmetric algorithm
Key management at centralized server	Agree	Agree	Agree	Symmetric algorithm
Group key management for cloud data storage	Disagree	Disagree	Disagree	Asymmetric algorithm

But they also need to worry about whether that huge information and their complete functionality are protective. But lots of complication arises while maintaining that information in a secure manner [22]. So, an effective privacy preservation strategy is needed in the field of cloud computing environment.

Existing research focus on overcoming various security attacks and enhancing various privacy function mechanisms. They also deal with the identification of different malicious and unethical attacks. They analysed how attack arises and which blocks oppose third parties in all aspects. Those techniques include identity management, cryptography, protocol-enhanced privacy mechanism, and so on [22]. So, the privacy preservation approach is introduced in the field of cloud computing environment. The privacy preservation approach deals with the overall potential and actual attacks from third parties in a cloud environment. It consists of various kinds of service and enhances security in a much improved and organized way. It also facilitates privacy protection embedded with the cloud computing environment (Figure 8.10).

Privacy protection plays a significant role in the field of cloud computing environment. Existing and proposed privacy protection mechanisms must wisely follow security. Various improved functionalities must be enhanced in the field of securing the data from an environment.

8.4.1 PRIVACY PRESERVATION AT CLOUD SERVICE SIDE

To maintain and provide privacy preservation is quite tricky at cloud service side. To overcome potential privacy risk, various privacy risks based on malicious threat from cloud users have been analysed and addressed using a privacy preservation mechanism [22].

8.4.1.1 Privacy Protection at Application Level of Cloud Service

Many existing approaches focus on protecting and enhancing privacy protection in the area of cloud computing environment. Smit et al. [23] discussed the entire framework and methodologies to overcome privacy policies in the cloud ecosystem. Based

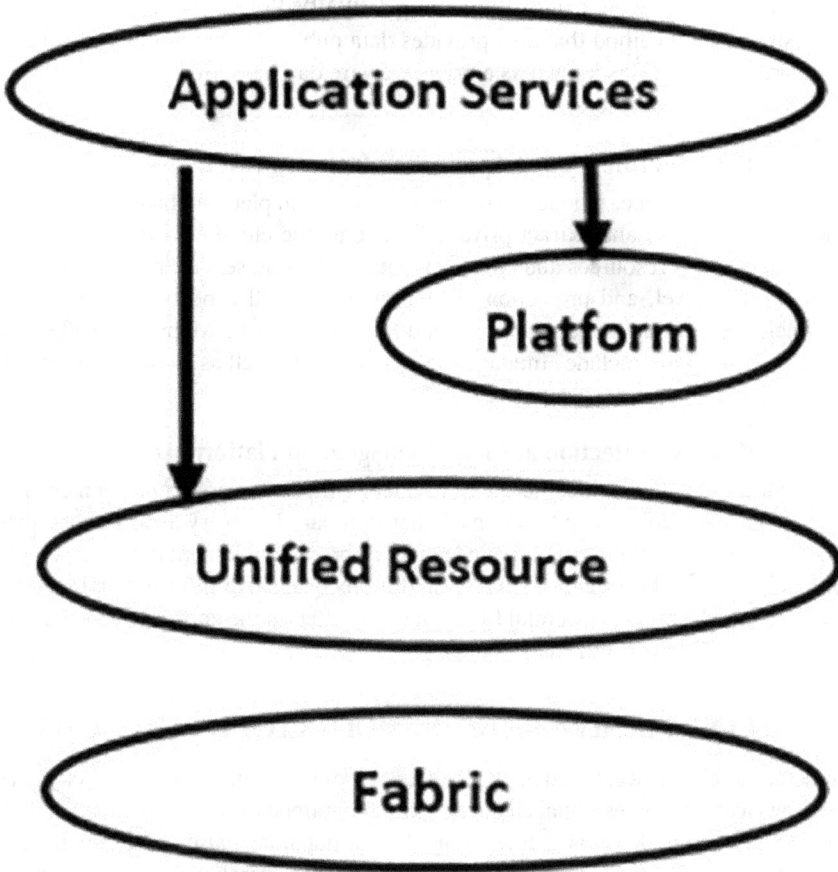

FIGURE 8.10 Cloud architecture.

on pollsters and respondents, a problem arises, known as the trust problem, by using mutual trustworthiness to achieve a typical semi-honest condition [24]. This technique can also rectify various limitations existing in a cloud environment. Mainly cloud environment has been deployed directly and indirectly.

We have to see how specific methods have been focused to rectify various privacy issues.

1. **Privacy-preserving data mining (PPDM)**: Suppose privacy leakage happens; it is revealed within a minute [25]. Evfimievski et al. [26] stated that, by using random operators, we have to extract patterns and analyse and discuss association rules [27].

2. **Privacy information retrieval (PIR)**: It helps keep the database operators and handles many exciting mechanisms using PIR. Beimel et al. [28] and Goldberg et al. [29] were involved in finding a theory known as deep dig in PIR. It enhances and intensifies the performance of various general methods [30].

3. **Privacy-preserving data publishing (PPDP)**: PPDP has been an extensively used method that also provides data publish web service [31]. SuLQ
framework [32] is known as a privacy-aware database; this helps to improvise the noise factor.

8.4.1.2 Privacy Protection at Application Platform Level

To support and service, a cloud environment application platform had been used, e.g.
Hadoop. To analyse and extract privacy functions, the cloud environment involves
accessing various resources and enhancements. In a cloud service provider, security
level, privacy level, and protection are maintained in all aspects. Privacy protection includes various preservation techniques consisting of MapReduce [33]. The
application program includes fundamental sequences as well as cloud customers and
enhances the development sequence.

8.4.1.3 Privacy Protection at Cloud Management Platform Level

In an entire sequence, cloud service providers enhance various management platforms to access step-by-step functions. When compared to every arrangement, these
preservation techniques might include and improve fundamental roles. A novel
approach needs to be applied to extract and protect sensitive information in a cloud
environment. A separate essential function is allocated in the entire sequence in such
a way as to avoid randomized avoidance.

8.5 RECENT VULNERABILITIES IN MULTI-CLOUD PARADIGMS

On-demand self-service, broad network access, resource pooling, rapid elasticity, measured services, etc. are essential characteristics maintained to efficiently function in the
cloud environment. A threat is like an incident of potential cause in a cloud environment and may affect the entire system or organization. A threat and its further exploration are termed as vulnerability. Taking more than one vulnerability is exploited by a
threat agent [34]. It includes various attacks and exposure in which different unauthorized functions also collapse the entire system (Table 8.3).

TABLE 8.3
Various Threats and Their Possible Vulnerabilities

Threats	Possible Vulnerabilities
Data breaches	1. Targeted attack
	2. Chance of getting rare human error
	3. Vulnerability in application
	4. Enhance low policies in security-wise
Loss of data	1. Natural hazards
	2. Chance of getting rare human error
	3. Failures of hard drive
	4. Power fluctuation
	5. Malware infection

(Continued)

TABLE 8.3 (*Continued*)
Various Threats and Their Possible Vulnerabilities

Threats	Possible Vulnerabilities
Malicious attacks	1. Network administrator 2. Contractor of third party 3. Various business parties 4. Former employees
Denial of service (DoS)	1. Very poor network architecture 2. Protocol network in insecure manner 3. Application of vulnerability
APIs and vulnerability	1. Management of key 2. Bugs of OS 3. Bugs of hypervisor 4. Software unpatched
Weak authentication and management of identity	1. Attack of social engineering 2. Man-in-the-middle attack 3. Malware infection
Account hijacking	1. Attack based on social engineering 2. Man-in-the-middle attack 3. Malware infection
Vulnerabilities in shared technology	1. Vulnerabilities on VM 2. Vulnerabilities on hypervisor 3. Vulnerabilities on third-party software
Lacking due diligence	1. Auditing is not maintained 2. Agreement in service level
Threats in advanced persistence	1. Spear phishing 2. Hacking in direct level 3. Malware in USB 4. Network penetration 5. Third-party applications
Cloud service abuse	1. Cloud service monitoring is not maintained 2. Agreement in service level
Poor responsibility	1. Negligence in human activities 2. Issues in service level
Security tools are insufficient	None
Human error	1. Security training is insufficient 2. Negligence in human activities
IoT device in unprotected way	1. Weak device management 2. Vulnerabilities in network 3. Vulnerabilities in hardware
Ransomware	1. Infrastructure vulnerabilities 2. Vulnerabilities based on platform 3. Vulnerabilities based on application

8.6   CLOUD DATA PROTECTION AND PRIVILEGE CONTROLS

To enhance security in cloud service platform, adequate data protection mechanism is required. Many malicious attackers, vulnerable in the cloud platform, have a chance to arise privacy leakage. Also, securing data in the cloud is a somewhat tricky task. To avoid that, various privacy protection mechanisms are involved in rectifying. But still, some inconsistency has occurred. So, we discuss and see some other practical tools to handle these problems and propose a new approach from the existing one.

8.6.1   Bɪᴛ Sᴘʟɪᴛ TᴇᴄʜɴɪQᴜᴇ

It is an effective encryption technique. When information is passed to the encryption mechanism, a random split occurs there. So, complete information would be split, and there is no chance of leakage. Then it provides an encryption mechanism in that output. In such a way, the encryption process is done. And also, pass this to the control function to provide controlling of sequence. Finally, the decryption mechanism involved decrypts the data from the encrypted data and the combining operation occurs (Figure 8.11).

Read data: 'Users' data are obtained in this state, in which all the data will be in raw format. Data from are like unencrypted data.

Split and shift: This process can split the data, in which the complete information is split up for user's convenience; after that, the shift process occurs on the split data. The data position is changed so that the entire data are shifted and look clumsy.

Encryption: The encryption process gets data from the split and shift processes. It enhances the data into an encrypted form to secure the user's information.

Recombining: It does the process of recombining the data. The output we get from this state is similar to that before the split and shift processes. But that data are in the encrypted format.

Decryption: Decryption gets the original information in the form of the user's information. These are some different strategies to do this process (Figure 8.12).

```
Algorithm to enable bit split technique:

Input: A set of File with accessible format.
Output: Data enabled with Bit Split Technique
for every request r in PQ do
If respaced exist in Trust database then
r.truth = true // Request is true and passed to split
else
truth_establishment(r) // Enable a trust relationship
end if
split = data / n; //n denotes user input i.e. [n=5]
result = split(r) // verify access roles in the request
if result = true then
shift = char[result] + 1; //It shifts every character after
1 digit
else
MS_Notify_User(r) // The MS notifies data owner
```

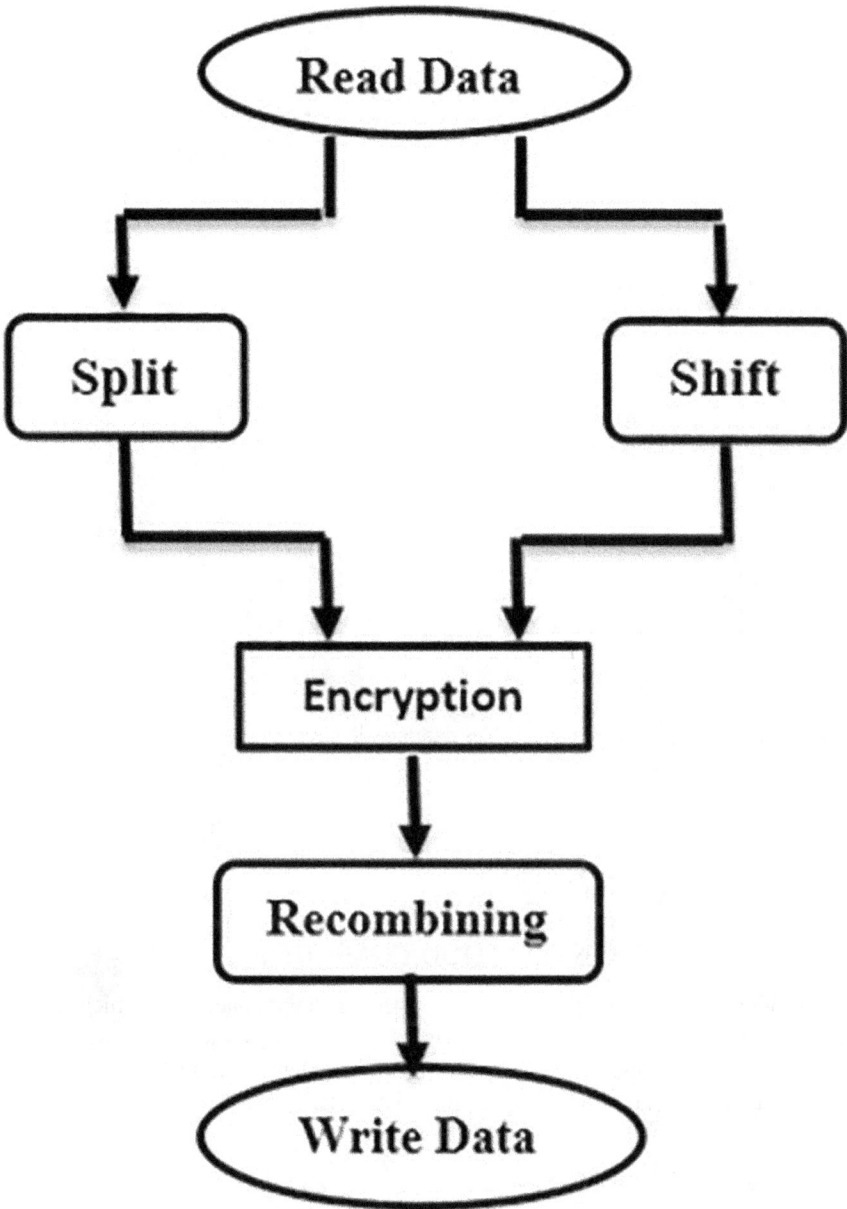

FIGURE 8.11 Bit split technique.

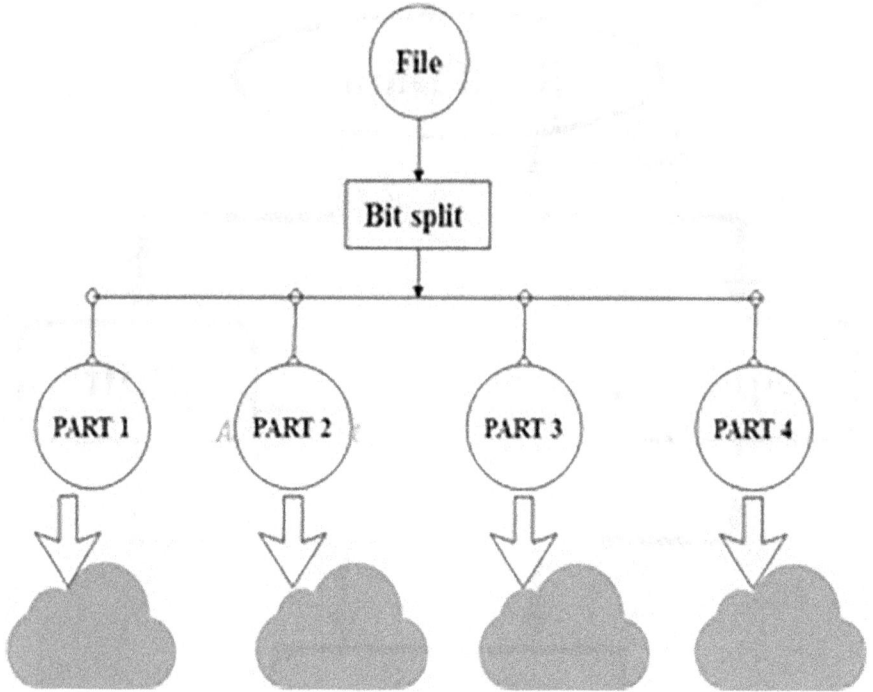

FIGURE 8.12 Bit split technique enables multiple clouds.

```
Final = result+shift;
end if
end for
```

8.7 SECURITY ISSUES OF CLOUD COMPUTING

On considering any technology, security is the primary concern in which a lot of research is still going to overcome various security issues in the cloud environment. Researchers have proposed many security enhancement mechanisms to overcome problems and different vulnerable and malicious attacks [41].

8.7.1 LACK OF ACCESS MANAGEMENT

In cloud computing environment, access management is the most critical issue. The access point is the primary factor for everything. That's the reason everyone targets it very much [41].

The primary reasons are:

1. Insufficient crisis management.
2. An information campaign that is very ineffective.
3. Very cunning hackers.

Most of the accounts were hijacked, and this causes the total drop of the network. To avoid this issue, multi-factor authentication is enhanced. It is the mechanism for providing security in a much enhanced way. Therefore, vulnerable threats have been identified and rectified at the initial stage so that network would be much stable.

8.7.2 DATA LEAK AND DATA BREACH

This is another main security concern in the cloud environment [41].

   a. **A data breach** is known as a leading cause and practical thing. If this happens, the entire confidential information will collapse and the whole environment gets inaccessible. It is like an accident in which complete information about the climate gets accessed without gets any authorization.
   b. **Data leak** is another primary security concern in which information leakage arises in the cloud environment. It paves the way for deletion of all the confidential information, which also has a chance that the overall environment may drop.

8.7.3 DATA LOSS

Data loss is the most threatening parameter in the process of communication environment on the cloud ecosystem. It is complicated to predict and difficult to handle considering various cloud environments. In such a way, it includes numerous times involved in it [41].

   a. **Alteration of data:** Usually, this issue happens in dynamic databases. Suppose a piece of information gets changed or altered. It does not support going back to the previous states.
   b. **Storage medium as unreliable:** Due to issues in the cloud's service provider, all the data get lost.
   c. **Deletion of data:** Once the information gets erased, we can't get back. Backup data may also get dropped.
   d. **Poor access:** Access to detailed information might get poor, so the random sequence mechanism is followed to satisfy access to the environment's particular function.

8.8   ADVANCEMENTS IN CLOUD SECURITY STRATEGIES

Today technology plays a vital role and also grows at a tremendous rate. Data accessibility has been a complicated thing, and storage is also a primary concern because massive amounts of data have been rendering. So, processing and managing have been considered as the topmost issues.

Hence, effective mechanism is needed to address and enhance the security strategies. It is based on the functionality of various sequences and numerous technologies are involved to make these functions in a much advanced way (Figure 8.13).

FIGURE 8.13 Various strategies involved in cloud security.

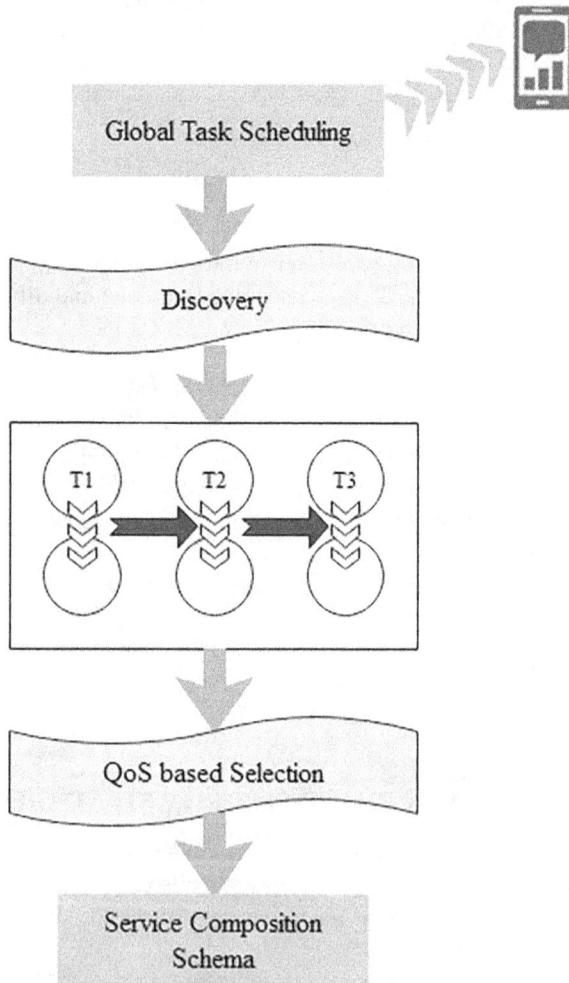

FIGURE 8.14 Sample privacy-aware data aggregation method.

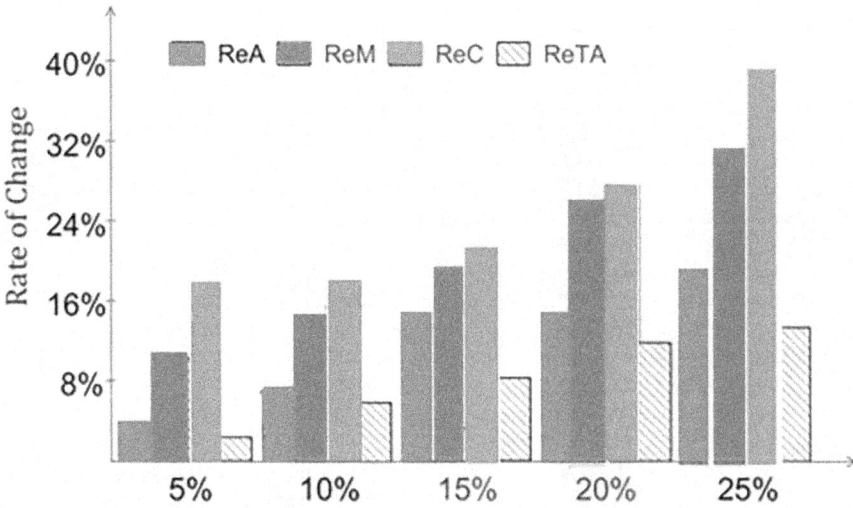

FIGURE 8.15 Comparison of different reputed mechanisms.

The cloud computing environment usage has become a routine and is of comparably low cost to access privacy control sequence. Hence, this paves the way to get all the necessary and fundamental information protected in a much more secure way. Figure 8.14 shows the overall privacy-aware data aggregation methods [43].

8.9 CONCLUSIONS

To improvise security in cloud environment, the existing methodologies of repute are used. For results comparison, various reputed mechanisms are involved in cloud security strategies, which use different rate parameters as shown in Figure 8.15. The parameter with mechanisms such as ReA, ReM, ReC, and ReTA with value of 25% also achieved best security functions. It makes cloud security environment much trustworthy and complex-free. Additionally, the assessment model of reputed parameters of ReC achieves the best accuracy in all aspects. Furthermore, it encounters all the upcoming possible attacks. Hence, the chance of malicious attack is very less and provides reliable services in all aspects in all possible mechanisms.

REFERENCES

1. Ghori, Muhammad Rizwan and Abdulghani Ghani. Review of access control mechanisms in cloud computing. *Journal of Physics: Conference Series*, 1049, 012092, 2018.
2. Karataş, Gözde and Akhan Akbulut. Survey on access control mechanisms in cloud computing. *Journal of Cyber Security and Mobility*, 7(3), 1–36, 2018.
3. Almubaddel, Majed, and Ahmed M. Elmogy. Cloud computing antecedents, challenges, and directions. *In Proceedings of the International Conference on Internet of Things and Cloud Computing*, p. 16, 2016.

4. Kumar, K. Suresh, T. Ananth Kumar, A. S. Radhamani, and S. Sundaresan. 3 blockchain technology. In: E. Golden Julie, J. Jesu Vedha Nayahi, Noor Zaman Jhanjhi (Eds), *Blockchain Technology: Fundamentals, Applications, and Case Studies*, p. 23. Boca Raton, FL: CRC Press, 2020.

5. Subashini, S. and V. Kavitha. A survey on security issues in service delivery models of cloud computing. *Journal of Network and Computer Applications*, 34(1), 1–11, 2011.

6. Punithasurya, K. and S. Jeba Priya. Analysis of different access control mechanism in cloud. *International Journal of Applied Information Systems*, 4(2), 34–39, 2012.

7. Xia, Y., L. Kuang, and M. Zhu. A hierarchical access control scheme in cloud using HHECC. *Information Technology Journal*, 9(8), 1598–1606, 2010.

8. Balasubramanian, S., S. Pratheep, R. Rajmohan, T. Ananth Kumar, and M. Pavithra. SVM block based neural learning technique for identification of fraudulent web pages. *Global Journal on Innovation, Opportunities and Challenges in Applied Artificial Intelligence and Machine Learning*, 4(2), 2020. ISSN: 2581-5156 (online).

9. Weber, Hazen A. Role based access control: The NIST solution. San Institute of Info Reading Room, October 3, 2008.

10. Park, Joon S., Gail-Joon Ahn, Ravi Sandhu. Role-based access control on the web using LDAP. In: M. S. Olivier and D. L. Spooner (Eds) *Database and Application Security XV. IFIP—The International Federation for Information Processing*, vol. 87. Boston, MA: Springer, 2002.

11. Yang, Caixia, Liang Tan, Na Shi, Bolei Xu, Yang Cao, and Keping Yu. AuthPrivacyChain: A blockchain-based access control framework with privacy protection in cloud. *IEEE Access*, 8, 70604–70615, 2020.

12. Sun, Pan Jun. Privacy protection and data security in cloud computing: A survey, challenges, and solutions. *IEEE Access*, 7, 147420–147452, 2019.

13. Chaudhry, Shehzad Ashraf, Khalid Yahya, Fadi Al-Turjman, and Ming-Hour Yang. A secure and reliable device access control scheme for IoT based sensor cloud systems. *IEEE Access*, 8, 139244–139254, 2020.

14. Rountree, Derrick and Ileana Castrillo. The basics of cloud computing: understanding the fundamentals of cloud computing in theory and practice. Waltham, MA: Syngress, 2014.

15. Dillon, Tharam, Chen Wu, and Elizabeth Chang. Cloud computing: Issues and challenges. *In 2010 24th IEEE International Conference on Advanced Information Networking and Applications*, pp. 27–33, IEEE, 2010.

16. Jing, Xue and Zhang Jian-Jun. A brief survey on the security model of cloud computing. *In 2010 Ninth International Symposium on Distributed Computing and Applications to Business, Engineering and Science*, pp. 475–478, IEEE, 2010.

17. Islam, Shareeful, Moussa Ouedraogo, Christos Kalloniatis, Haralambos Mouratidis, and Stefanos Gritzalis. Assurance of security and privacy requirements for cloud deployment models. *IEEE Transactions on Cloud Computing*, 6(2), 387–400, 2015.

18. Qiu, Yuqin, Qingni Shen, Yang Luo, Cong Li, and Zhonghai Wu. A secure virtual machine deployment strategy to reduce co-residency in cloud. *In 2017 IEEE Trustcom/ BigDataSE/ICESS*, pp. 347–354, IEEE, 2017.

19. Che, Jianhua, Yamin Duan, Tao Zhang, and Jie Fan. Study on the security models and strategies of cloud computing. *Procedia Engineering*, 23, 586–593, 2011.

20. Tysowski, Piotr K. and M. Anwarul Hasan. Hybrid attribute-and re-encryption-based key management for secure and scalable mobile applications in clouds. *IEEE Transactions on Cloud Computing*, 1(2), 172–186, 2013.

21. Lei, Sun, Dai Zishan, and Guo Jindi. Research on key management infrastructure in cloud computing environment. *In 2010 Ninth International Conference on Grid and Cloud Computing*, pp. 404–407, IEEE, 2010.

22. Zhang, Gaofeng, Yun Yang, Xuyun Zhang, Chang Liu, and Jinjun Chen. Key research issues for privacy protection and preservation in cloud computing. *In 2012 Second International Conference on Cloud and Green Computing*, pp. 47–54, IEEE, 2012.
23. Smit, Michael, Kelly Lyons, Michael McAllister, and Jacob Slonim. Detecting privacy infractions in applications: A framework and methodology. *IEEE 6th International Conference on Mobile Adhoc and Sensor Systems (MASS'09)*, pp. 694–701, Macau, China, October 12–15, 2009.
24. Golle, Philippe, Frank McSherry, and Ilya Mironov. Data collection with self-enforcing privacy. *ACM Transactions on Information and System Security*, 12(2), 1–24, 2008.
25. Agrawal, Rakesh and Ramakrishnan Srikant. Privacy-preserving data mining. *ACM SIGMOD Record*, 29(2), 439–450, 2000.
26. Evfimievski, Alexandre, Johannes Gehrke, and Ramakrishnan Srikant. Limiting privacy breaches in privacy preserving data mining. *In 22nd ACM SIGMOD-SIGACT-SIGART Symposium on Principles of Database Systems (PODS 2003)*, pp. 211–222, San Diego, CA, June 09–11 2003.
27. Liu, Li, Murat Kantarcioglu, and Bhavani Thuraisingham. The applicability of the perturbation based privacy preserving data mining for real-world data. *Data and Knowledge Engineering*, 65(1), 5–21, 2008.
28. Beimel, Amos, Yuval Ishai, and Eyal Kushilevitz. General constructions for information-theoretic private information retrieval. *Journal of Computer System Science*, 71(2), 213–247, 2005.
29. Goldberg, Ian. Improving the robustness of private information retrieval. *2007 IEEE Symposium on Security and Privacy (SP'07)*, pp. 131–148, Oakland, CA, May 20–23, 2007.
30. Henry, Ryan, Femi Olumofin, and Ian Goldberg, Practical PIR for electronic commerce. *18th ACM Conference on Computer and Communications Security (CCS'11)*, pp. 677–690, Chicago, IL, October 17–21, 2011.
31. Fung, Benjamin C. M., Ke Wang, Rui Chen, and Philip S. Yu. Privacy-preserving data publishing: A survey of recent developments. *ACM Computing Surveys*, 42(4), 1–53, 2010.
32. Blum, Avrim, Cynthia Dwork, Frank McSherry, and Kobbi Nissim. Practical privacy: The SuLQ framework. *24th ACM SIGMOD-SIGACT-SIGART Symposium on Principles of Database Systems (PODS 2005)*, pp. 128–138, Baltimore, MD, June 13–16, 2005.
33. Rastogi, V., Dan Suciu, and Sungho Hong. The boundary between privacy and uitility in data publishing. *Proceedings of the 33rd International Conference on Very Large Data Bases (VLDB 2007)*, pp. 531–542, Vienna, Austria, September 23–27, 2007.
34. Suryateja, P. S. Threats and vulnerabilities of cloud computing: A review. *International Journal of Computer Sciences and Engineering*, 6(3), 297–302, 2018.
35. Zhang, Wei, Xinwei Sun, and Tao Xu. Data privacy protection using multiple cloud storages. *In Proceedings 2013 International Conference on Mechatronic Sciences, Electric Engineering and Computer (MEC)*, pp. 1768–1772, IEEE, 2013.
36. Dang, Thanh Dat and Doan Hoang. A data protection model for fog computing. *In 2017 Second International Conference on Fog and Mobile Edge Computing (FMEC)*, pp. 32–38, IEEE, 2017.
37. Colombo, Maurizio, Rasool Asal, Quang Hieu Hieu, Fadi Ali El-Moussa, Ali Sajjad, and Theo Dimitrakos. Data protection as a service in the multi-cloud environment. *In 2019 IEEE 12th International Conference on Cloud Computing (CLOUD)*, pp. 81–85, IEEE, 2019.
38. Ahmadi, Mohammad, Faraz Fatemi Moghaddam, Amid Jamshidi Jam, Somayyeh Gholizadeh, and Mohammad Eslami. A 3-level re-encryption model to ensure data protection in cloud computing environments. *In 2014 IEEE Conference on Systems, Process and Control (ICSPC 2014)*, pp. 36–40, IEEE, 2014.

39. Lang, Bo, Jinmiao Wang, and Yanxi Liu. Achieving flexible and self-contained data protection in cloud computing. *IEEE Access,* 5, 1510–1523, 2017.
40. Hashizume, Keiko, David G. Rosado, Eduardo Fernández-Medina, and Eduardo B. Fernandez. An analysis of security issues for cloud computing. *Journal of Internet Services and Applications,* 4(1), 5, 2013.
41. Kumar, K. Suresh, T. Ananth Kumar, A. S. Radhamani, and S. Sundaresan. Blockchain technology: An insight into architecture, use cases, and its application with industrial IoT and big data. In: E. Golden Julie, J. Jesu Vedha Nayahi, Noor Zaman Jhanjhi (Eds) *Blockchain Technology: Fundamentals, Applications, and Case Studies,* pp. 23–42. Boca Raton, FL: CRC Press, 2020.
42. Olushola, Omoyiola Bayo. Strategies for securing cloud services, 2020.
43. Wen, Yiping, Jianxun Liu, Wanchun Dou, Xiaolong Xu, Buqing Cao, and Jinjun Chen. Scheduling workflows with privacy protection constraints for big data applications on cloud. *Future Generation Computer Systems,* 108, 1084–1091, 2020.

9 Quantum Computing and Quantum Cryptography

B. Sheik Mohamed
EY LLP

M. Satheesh Kumar and K.G. Srinivasagan
National Engineering College

CONTENTS

9.1 INTRODUCTION

In the world of this digital era, everything is controlled by an electronic device called computer. Each year, the computational power is growing exponentially due to the increase in the demand for complex computations. Whenever we solve a mystery in the universe, we open the door to an unknown complex problem. In the famous Apollo mission, the moon landing programme was achieved with the help of a computer, which had a memory of 15 bit word length + 1 bit parity, 2048 words RAM. Now, we have exponentially greater power in handheld devices such as smartphones in comparison with the Apollo guidance computer. This rapid improvement is because we need to process the new data which we find from the complex problems, and the engineers build more powerful machines [1] year by year. But there is a limitation for even the fastest powerful supercomputers which cannot solve or simulate the solution for many unknown problems. In olden days, people used candles to bring light into their houses by igniting them. It can produce a limited amount of light, and once the candle dies, people spent their times in dark. In that candle era, scientists and engineers designed the electric bulb. Both the candle and the bulb are used for generating light, but they are not the same. A quantum computer is not exactly a superdominant form of our traditional digital computing machines; it is just like the comparison we mentioned – an electric bulb is not exactly a powerful candle. It is impossible to design an electric light bulb by creating best and superior candles, because the electric bulb is built on completely superior technology which stands on the deeper physics concept called electromagnetism. Correspondingly, a quantum computer is a novelty of computing machine [2] that stands on the deeper principles of quantum physics. In the same way, with the traditional computers we have certain limitations in processing complex data. To overcome this problem, physicists started working on the quantum properties [3] to construct a machine to accomplish the monumental tasks that cannot be processed by traditional computers (including supercomputers). This machine is known as quantum computer. In comparison, it can be an electric bulb to the candle scenario where quantum computers can outperform traditional computers with a giant leap. To understand the concept of quantum computing [4], we must have a fundamental knowledge on the following:

- How quantum physics came? (origin of modern physics)
- What is wave and particle nature (duality nature of light) of sub-atomic particles?
- Concepts of ultraviolet catastrophe and blackbody radiation
- Concepts of qubits
- Concepts of quantum entanglement and superposition.

The objective of this chapter is to provide a comprehensive and complete overview of quantum cryptography and to offer a comprehensive analysis of quantum computing. The rest of this chapter is structured as follows: Section 9.2 contains the basic nature of light. In Section 9.3, we provide the origin of modern physics. In Section 9.4, we describe the applications of quantum physics where we offer a complete explanation of quantum computing. In Section 9.5, we discuss the detailed note on quantum cryptography. Section 9.6 summarizes the chapter with the future of quantum.

9.2   NATURE OF LIGHT

Before 1900, the era of classical mechanics, physicists were aware of light as a wave and no one understood the dual nature of light. Yes, light behaves like both wave and particle. In the Newtonian era, there was a great debate between physicists on light as a wave or particle. Sir Isaac Newton put forth the idea that light is made up of particles, and Christiaan Huygens suggested the wave theory. Both sides argued how the phenomena such as reflection, refraction, and diffraction take place. Due to Newton's popularity on that era, Huygens idea of light as a wave was not widely accepted. After Young's double-slit experiment, as depicted in Figure 9.1, physics community accepted that the light is a wave. The double-slit experiment can be easily understood

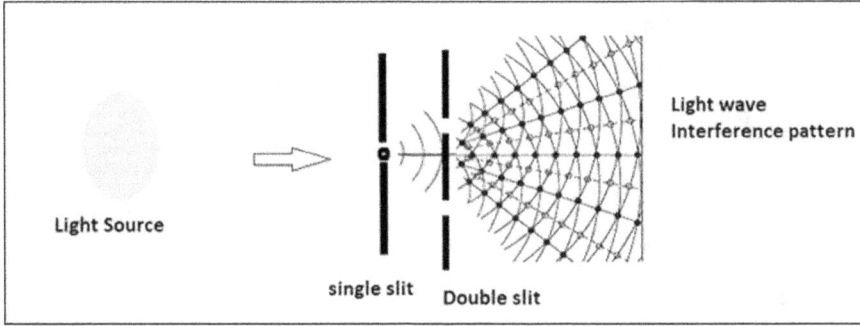

FIGURE 9.1 Graphical representation of interference pattern caused by light as a wave using the double-slit experiment.

in that the interference pattern [5] is formed when the light travels via double slit as a wave.

9.3 ORIGIN OF MODERN PHYSICS

Physics community has observed exponential growth in the last 120 years, where modern physics came into existence with two strong pillars called relativity and quantum mechanics. Back in 1900s, the budding moment of modern physics [6], remarkable physicists such as Max Planck and Albert Einstein contributed to the new field of physics called quantum mechanics or quantum physics [7]. The thought of light as a wave raised a new problem in that time, ultraviolet catastrophe. To understand the ultraviolet catastrophe, one must know the concept of an ideal blackbody. An ideal blackbody absorbs all the electromagnetic radiation (remember light is a form of electromagnetic radiation) irrespective of the frequencies. As same as absorption, the blackbody emits the thermal electromagnetic radiation. So, as per classical physics, it leads to ultraviolet catastrophe. Using classical mechanics foundations, physicists explain the blackbody radiation in terms of frequency and the corresponding temperature. English Physicists, Lord Rayleigh and Sir James Jeans derived the law known as Rayleigh–Jeans law, using classical foundations as depicted in Figure 9.2. According to this law, output energy from shorter wavelength would result in infinity. During observations, it is different. Such difference in the observation and mathematical prediction is known as the ultraviolet catastrophe.

At the same time, the famous German Physicist, Max Planck solved this catastrophe proposing light as energy quanta. It is the breakthrough in the physics field that led to the birth of the new pillar of modern physics called quantum physics [8]. It leads to the dual nature of light as wave and particle. We highlighted the full list of notable works in the field of quantum physics (since 1900) in Table 9.1.

In 1905, Albert Einstein published the paper on the photoelectric effect; eventually, for his study on photoelectric effects, Einstein won Nobel Prize in 1921. Physicists of that era started to work on the behaviour of wave and particle nature of photons and other sub-atomic particles such as electron. Scientists such as Niels Bohr, Erwin

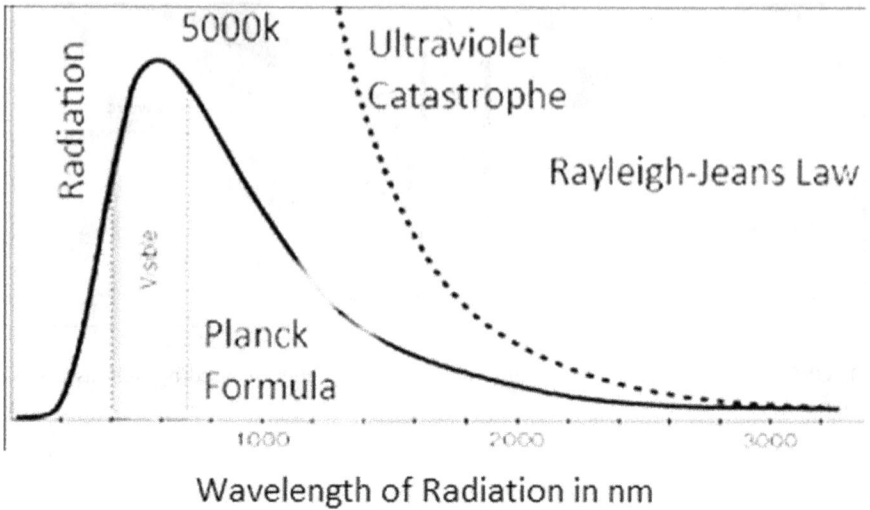

FIGURE 9.2 Graphical representation of ultraviolet catastrophe using Rayleigh–Jeans law and observational consistency using Planck's formula.

TABLE 9.1
Notable Discoveries in Quantum Physics

Year	Notable Discoveries and Experiments	Physicist
1900	Blackbody radiation	Max Planck
1905	Photoelectric effect	Albert Einstein
1911	Atomic nucleus discovery	Ernest Rutherford
1913	Atomic hydrogen model	Niels Bohr
1923	Compton scattering	Arthur Holly Compton
	De Broglie hypothesis (proposed all matter has wave properties)	Louis de Broglie
1925	Dynamical pictures (Heisenberg picture and Schrödinger picture)	Werner Heisenberg
		Erwin Schrödinger

Schrödinger, Werner Heisenberg, and Paul Dirac gave the proper understanding of the concept of quantum physics. The most famous one is Schrödinger's cat in the box experiment. It was a thought experiment where a cat was placed inside the box with a small amount of radioactive source, a flask of hydrocyanic acid (poison), and a Geiger counter device (to detect if any decay of atom). If Geiger counter detects any decay, they positioned the hammer to shatter the flask containing poison. In case of shattering, the flask will release the poison and the cat inside the box is dead. Unless we observe the state of the cat, the probability says that the cat is being alive and dead. When you open the box, if the cat is found alive, the observer remembers it is being alive, or vice versa. This explains the concept of superposition [9] that the cat is being alive and dead until no observation happens. This phenomenon of superposition happens in sub-atomic particles.

9.4 APPLICATION OF QUANTUM PHYSICS

9.4.1 QUANTUM COMPUTERS

Physics has always come up with real-world applications, to provide the solution to an unknown problem. Yes, we heard it correctly, unknown problem. Humankind has never seen several things as a problem in the past, but physicists vehemently solved them. Their mathematical ideas allowed engineers to build monuments, ships, air-crafts, motor engines (including all the vehicles we are using now), electricity, elec-trical devices (such as TV, radio, watch, and refrigerators), and many more. All these applications usually attributed to a general branch of physics [10], and the question always comes up on applications of quantum mechanics. Indeed, it has a huge number of real-world applications such as computing, drugs making, molecule processing, and a wide range of simulations in particle and quantum physics, which improve the lives of millions of people. In the space of computing, physicists applied the quantum properties for the functioning of logical manipulation and computing; it is popularly known as quantum computers [11]. First, we need to understand the function of the traditional computer. The invention of computers reduced the human efforts in expo-nential scales. The computer itself is a real-world application of physics.

Computers perform the task with the help of integrated circuits as described in Figure 9.3. The circuits are made up of logic gates, and they are working with the support of transistors. In all the electronic devices, the role of transistors is to con-trol the flow of electron either by allowing or stopping the flow with the help of a switch. If the electron is allowed, then it is ON state and if stopped it is OFF state. We understand this '0' (OFF state) and '1' (ON state) as bits. Using logic gates, these bits are used for desired calculations. By assigning more logic gates, we ideally use it for solving complex problems such as satellite trajectory, space probing, and modern-day usages such as maintaining documents and sending emails for busi-ness functions. The exponential increase in the processing power of the chips in the contemporary world has led to the usage of computers literally in every field. In fact, semiconductors are the building blocks of electronic gadgets which use the proper-ties of quantum mechanics only. It is achieved by the process called doping, which adds impurities in the conducting material to alter its properties of conducting; this

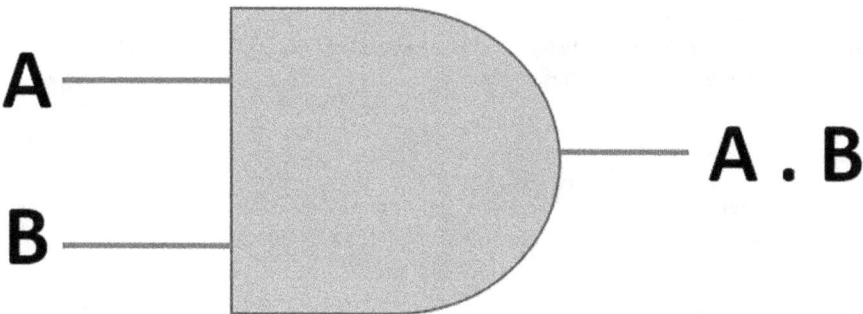

FIGURE 9.3   Integrated circuits.

is where the junction is created in the semiconductor. In most of the modern components in electronics, such as transistors, diodes, and logic gates, the junction only provides the core functionality of those components. So, can we refer to the digital computers as quantum computers [12], or do we have any other criteria to be referred to as quantum computing machines?

Well, the digital computers, lasers, telecommunications, and electronic gadgets are the real-world applications of quantum physics. But they are not quantum computers. Why?

In the current era, computer makers beef up the processing power of computers by scaling more chips and cores to provide the faster processing. It may be playing ultra-definition videos or high-end graphical games, which usually require extreme processing power. If we are able to increase the processing power, then why big giants such as IBM, Microsoft, Google, Amazon, and Alibaba [13] are investing billions in building the quantum computers.

9.4.2 Quantum Computing

In a traditional computing device, things are acted up in binary, i.e. zeros and ones. As we discussed, we call it 'OFF' and 'ON' states. In quantum computing, we are not going to deal with traditional bits like binary, but we are going to deal with the concept called 'qubit' [14].

9.4.3 Qubit

Qubit is a quantum bit like the traditional binary system in the digital computers, the bits '0' and '1' in superposition. So, qubit is also referred to as the fundamental block of information in quantum computers. Elementary particles, well known as sub-atomic particles that include photons or electrons, can be utilized in a quantum device similar to the serving polarization or its charges as a resemblance of zero and/or one. These particles are now referred to as a qubit [15]. Using the quantum property, these particles are implied to express the quantum theory and perform the computing, known as quantum computing. In quantum physics, there are two important principles such as superposition and entanglement which will be used for quantum computing. Superposition and quantum entanglement are the most mind-boggling concepts ever encountered in reality by not just layman but even great scientists such as Albert Einstein. Let us understand the concepts of superposition and entanglement in a brief description.

9.4.4 Superposition

In the earlier section, we discussed the famous thought experiment of Erwin Schrödinger, a cat in the box experiment, which describes the superposition [16] of cat in alive and dead state at the same time until someone observes. Let us assume a similar scenario in which an electron is placed in a magnetic field. The spin of the electron can be expressed in two states as a spin-up state and a spin-down state (both are opposite states like zero and one). Moving the spin of the electron from one state to another (spin-up/spin-down) that is one by using an energy beam, similarly like a laser beam – need to assume we use 1 laser energy unit. Assume that just 1/2-unit laser energy is used and particles are entirely separated from any other effects. As per the quantum mechanics principles [17], the sub-atomic particle then goes into a superposition state, in which it is in both the states at the same instance. In this way, every qubit can take a superposition of both 0 and 1. Therefore, the quantum computer can perform 2^n (number of calculations), where n is the number of qubits used. In the par, if a quantum computer with qubits of 500 which had the probability of performing 2^{500} computations in one go. According to astrophysicists, the number 2^{500} could exceed the number of atoms in the observable universe. In order to make the particles interact with one another, we need to understand the concept called quantum entanglement.

9.4.5 ENTANGLEMENT

In 1925, dynamical pictures of quantum mechanics were created by Werner Heisenberg (Heisenberg picture) and Erwin Schrödinger (Schrödinger picture). In the superposition, we discussed the Schrödinger picture, the thought experiment of cat in the box. To understand the concept of quantum entanglement [18], we must understand the concept called Heisenberg's uncertainty principle. Simply put, the concept of uncertainty states that the orientation and speed of a component cannot be measured at the very same time. Suppose we measure the velocity of the sub-atomic particle, then we cannot measure its position, and if we measure the exact position, then we cannot measure the velocity. It is hard to grasp since it appears to be non-logical. But in the scale of quantum level, the sub-atomic particles behave like this. It is because of the dual nature of the sub-atomic particles. As per the principle, the formula is:

$$\Delta x \Delta p \geq \frac{h}{4\pi} \tag{9.1}$$

$$\Delta x \Delta p \geq \frac{\hbar}{2} \tag{9.2}$$

where
  Δx: the uncertainty in position
  Δp: the uncertainty of momentum

  $\hbar$: $\frac{h}{4\pi}$ [where h is the Planck's constant ($6.62607015 \times 10^{-34}$ J · s)]
  π: pi (~3.14).

For position and momentum, the uncertainty principle can be explained with this equation. So, we understand the idea behind the uncertainty concept. But where the quantum entanglement arises here?

Albert Einstein and Niels Bohr were known for their personal level friendship. According to many sources, they were in fact the best friends. In terms of quantum mechanics, they behaved like nemesis. This is because Bohr was a strong supporter of quantum theory and Einstein was the main opponent. Even during one of the discussions, Einstein challenged Bohr [19] notoriously by stating, 'Would God play the world dice?' to which Bohr retorted wonderfully, 'Chose not to tell God what he would and can't do.' Quantum theory made Einstein so uneasy that he conceived an experiment to eliminate something after the study. But Bohr always came out triumphant. Drawing on their relevance to scientific philosophy, these discussions are recalled. Bohr appreciated Einstein, and Bohr's knowledge was revered by Einstein. In the early part of the 20th century, the two prominent physicists played a leading role in the study. In one such instance, Albert Einstein and Niels Bohr had a great debate on the quantum theory. Einstein brought up a new problem to disprove the quantum theory, and Bohr refuted it with his proof. In one such situation, the quantum entanglement was born. Quantum entanglement is always referred to as one of the complicated ideas in physics. Entanglement is perceived as a phenomenon uniquely quantum mechanical, but it is not. Contemplating a basic classical variant of entanglement first is enlightening, albeit somewhat unorthodox. This encourages us, aside from the general oddity of quantum theory, to interfere away the subtlety of entanglement itself. Sub-atomic particles that have interacted at some point maintain a form of association and may be intertwined as a pair in a mechanism referred to as correlation. Getting the knowledge of the spin state for one of the entangled particles, which may be in the position of down or up, instantaneously reveals another entangled particle's spin state, because one of the pair will be in the opposite direction since the first one is observed. Perhaps, further striking is the observation that, owing to the occurrence of superposition, the observed particle does not have a single direction of rotation until being calculated, but is instantaneously both spin-up and spin-down. The spin status of the computed component is now computed and passed to the entangled particle, which concurrently takes the exact opposite direction of rotation to that of the measured particle, even if the particles are separated by the distance of a light year. Einstein famously quoted this phenomenon as 'spooky action at a distance' [20]. The initial goal of Einstein was to measure the momentum of one particle and measure the velocity of the other particle to break the uncertainty principle. Bohr and Einstein agreed that this measurement would affect the state of the entangled particle. Really, it is a spooky action at a distance. The above phenomena allow qubits isolated from each other, particularly if they are put in distant universes, to converse instantly. Regardless of how far the connected particles are from one other, they stay intricate if separated. With the help of quantum entanglement and superposition [21], we might create an exceptionally enriched computing capability. In the traditional computer, a two-bit register can store only one of the four binary combinations like the following in any given time: 00, 01, 10, or 11.

A double-qubit record [22] could concurrently retain all four numbers on the quantum computer, since each qubit has two values. If we add additional qubits, the capacity is exponentially enhanced.

9.4.6 Popular Real-Life Scenario

Researchers set up a coin flipping game to simulate the probability of winning between traditional computer and human player as illustrated in Figure 9.4. The

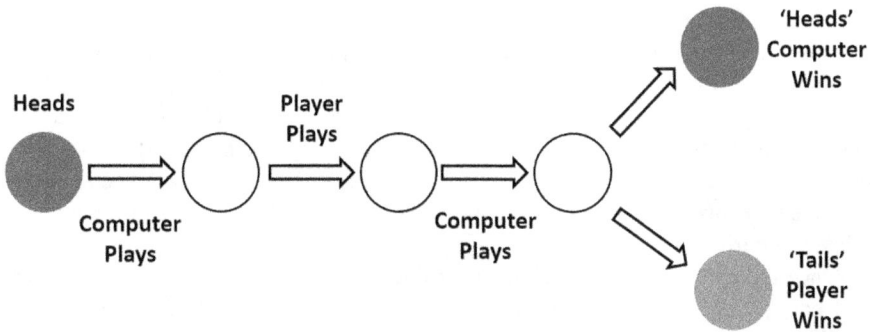

FIGURE 9.4 Coin Flipping game in traditional computer.

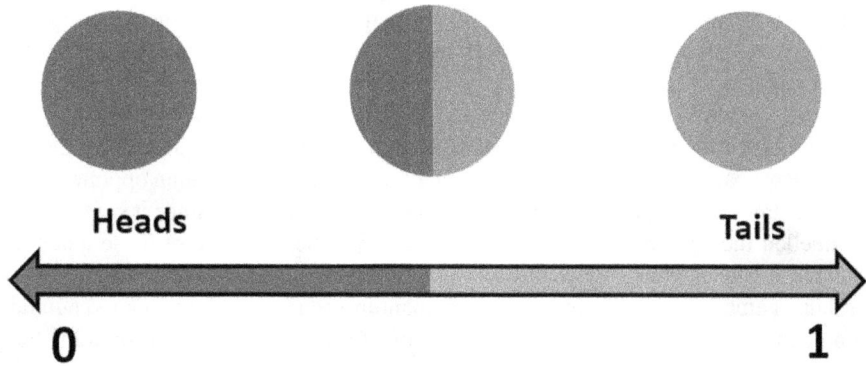

FIGURE 9.5 Coin Flipping game in quantum computer.

game is simple in that the coin shows heads and the machine [23] moves first by flipping or unchanging the side of the coin. Then the player must find the outcome without knowing what the machine did. When the human player's turn comes, you are allowed to perform the coin flip or not. Now the computer does not know the outcome of the player's move. At last, the machine's turn comes, and it might perform the coin flip or not. Once the three rounds are completed, the state of coin is shown finally. Suppose the coin results in 'heads', then the machine wins, and vice versa.

In this flipping game scenario [24], researchers noted that the human player got 53% of chance to win the game and the computer got the chance of 47% only. Fairly, we can say that the rate of winning might reach up to 50% (or around it), as we calculated in Figure 9.5. In the same game, if introduce quantum computers to play the game in the place of traditional computer, then the result is one-sided. The big question that comes to our mind is how it accomplishes this astonishing winning streak or is it any sort of double-dealing, but in reality, it is due to the nature of quantum physics.

Here is in what way it works. The traditional computer considers heads 'H' or tails 'T' of a coin as a bit, a '0' or a '1'; in digital, it is considered as OFF or ON state inside the machine's memory. In the quantum computer, it is totally distinctive. One qubit contains an extra form, nonbinary character. It can occur as a superposition; that is, it can be a mixture of 0 and 1, with chances of occurring zero and few probabilities of occurring one. The other way is to describe the range of its character: like a 70% occurrence of being zero (heads) and a 30% occurrence of being one (tails) or 40–60 or 65–35. These opportunities can be infinite. The considerable evidence seems to indicate that this could eventually wind up with zero and one result represents and allow insecurity.

Once the game happens, such a permeable combination of heads and tails, zero and one, produces a quantum computer, irrespective of the players' choice such as flipping or not flipping, the concept superposition comes into the picture, and it remains unharmed [25]. It is sort of similar to mingling a combination of different liquids. Mixing the liquids has no impact, as the combined liquid stays the same once the machine has closed. In the closing stage, the quantum computer will perform an operation on the two states and flawlessly undo the combination of the zero and one, restoring the game so that the player is always in a difficult situation. In the same coin flipping game conducted by IBM in their quantum computer, the result was 97% of the time the quantum computer defeated the human opponent and only 3% was the probability for the human player to win. Researchers from IBM quoted that the 3% winning probability is due to the operational failure, otherwise human opponent cannot gain any victory. The quantum computer emerged victorious since it successfully channelled the superposition and uncertainty [26]. The influence of these quantum features is far-reaching, not just for picking the right coin flip or drawing the best card in a game, but for building the future quantum technology. It has a good number of potential real-life applications that might transform our lives, such as building better life-saving drugs. Since describing and computing all the quantum characteristics of all the atoms in the molecule is a computationally unfeasible process, supercomputers are out of the question for doing so. As the current world is in the middle of pandemic, we understand two things:

- How important is the role of medicine in the pandemic situation?
- How difficult is it to create drugs for novel virus infection?

In future, quantum computers will help us produce more accurate results, because the system functions on the same quantum characteristics as the molecule it is trying to emulate. In the upcoming decades, a grand-scale quantum simulation for drug manufacturing might open the door for the treatment of diseases such as Alzheimer's, which damages the lives of thousands of elder populations all over the globe.

In the field of computing, we are always facing crisis in the breaking of computer security by hackers since the budding of digital computer. For securing the cyber world, quantum computing comes to play a huge role using quantum properties to build unbreakable cryptographic algorithms, which is called quantum cryptography [27]. The same form of strong encryption has already been put to test by a set of economic institutions across the world. It is estimated that currently there are over 17 billion connected devices around the world.

9.5   QUANTUM CRYPTOGRAPHY

Since 1970, the public key cryptography has enhanced and acted as a crucial factor for the worldwide digital communication network. A huge number of applications directly rely on the digital communication networks that are so vital to the global finance, defence, and control of the people's daily survival, such as cell phones, social media, e-commerce, and cloud computing. In the tangled world, the capacity of people, industries, and authorities to connect securely is of the highest priority. Most of the several important communication rules support predominantly with the following essential cryptographic routines (totally three routines):

- Public key encryption
- Digital signatures
- Key exchange.

Presently, the above-mentioned routines are largely applied with the help of RSA (Rivest–Shamir–Adleman) cryptosystem, elliptic-curve cryptosystems, and Diffie–Hellman key exchange [28]. The strength of these procedures depends on the number of theoretical concerns, such as polynomial interpolation or the continuous logging problem, that are presented across different parties. During the mid-1990s, Peter Shor, an innovator from Bell Labs discovered new quantum computers, a unique eco-system optimizing the physical characteristics of energy and matter to provide computational power, which can confidently deal with many of these abnormalities and therefore render all public key cryptographic algorithms based on these presumptions vulnerable. As a result, a massively good quantum computer will put various types of modern secure communication, such as secure key exchange, secure encryption, and digital authentication, at risk. With the revelation that quantum computers would likely be able to solve a variety of problems far quicker than conventional computers, a huge amount of interest has been sparked in quantum computing. Over the past two decades, Shor's discovery has revolutionized the field of quantum algorithms. And since then, the field of quantum algorithms has continued to grow. While it is true that we are only now learning about quantum algorithms' enormous efficiency, such a thing has been realized for a lot of complications in quantum and particle physics simulation (which we have seen in LHC – Large Hadron Collider), number hypothesis, and regional anatomy. Despite this, the list of issues that must be resolved in order to increase the rate of exponential speedup for quantum processing remains surprisingly small. In contrast, in order to pursue even faster progress in pursuing broad topics linked with probing, impact, and the discovery of crash-landed aircraft, an assortment of other techniques has also been put in place. Grosser's search method expands the linear speedup to be applicable to unstructured search situations. While an increase in the speed at which cryptographic algorithms run has no influence on the cryptographic capabilities of such algorithms, it can force higher key sizes in the symmetric key setting. Shor's algorithm is a premier example of quantum algorithm that shows the power of quantum computation compared to classical computation today. Shor's technique [29] finds prime factorization of huge numbers roughly exponentially faster than the fastest known classical algorithms. Finding factors for large numbers is the basis of RSA encryption, which is a public key sharing scheme used for secured data transactions, such as credit card transactions. One popular public key encryption scheme is RSA encryption, which enables communications to be classified with a passcode called a global password which may be freely shared. Once a communication has been encrypted using the public key, it can only be decrypted with the corresponding private key, which is a very large number. It can only be deciphered by a secret key, which is the prime factor of that large number. The larger the number, the more computational power it takes to find a prime factor and the more secure the encryption is. So, the defence of RSA depends on the statistic that discovering a prime factor of a large number takes an extremely long amount of time. For example, of a 232 decimal digit semi-prime number. A semi-prime number is just the product of two prime numbers, and it is the hardest type of number to factorize. To factorize the number, we need around a 1000 years of processing time on a typical computer. Shor's algorithm is a quantum algorithm for finding the prime factors of an integer N (it should not be a prime/even/integer power of a prime number).

For instance, check the below details on breaking of traditional cryptographic algorithm by Shor's algorithm using quantum computers: To crack into a cryptosystem and which have a preceding experience of one fact regarding N (the RSA public key): that N has precisely one prime factorization. So, in order to locate N's factors to crack the RSA encryption, first we need to evaluate $f(x) = (x^r)$ mod N. However, in this case, N is a humungous number, and instead of just guessing the factors, the most efficient, exact approach to obtaining 'r' is to calculate it using the hint that N is prime-factorizable and 'r' is periodic, i.e. $f(x) = f(x+r)$. In order to calculate the 'r' for a $f(x)$, Fourier transformations on periodic functions are utilized. The fundamental element of Shor's algorithm is to obtain at this location. This process involves a process known as the quantum discrete Fourier transform (QFT) [30] to discover the time 'r'. Basically, the Fourier transform is applied in several areas such as signal processing and data compression in the traditional computing field, whereas the quantum computing utilizes the enhanced Fourier transform, a technology that first emerged during the 19th century. The discrete Fourier transform is applied to the amplitudes of a wave function to compute the Fourier transforms of amplitudes, which are themselves the quantum application of the discrete Fourier transform. With the discrete Fourier transform, one takes a series of figures and converts them into a collection of sines and cosines. Rather than creating a list of the probabilities for the given set of qubit states, the QFT compiles a list of the 'probability amplitude' for the provided set of qubit states.

A quantum computer accomplishes this mission efficiently because of the subsequent motives:

1. The fact that it can operate in all possible states simultaneously (that is indisputable rapid processing) greatly enhances a quantum computer's ability to gauge the periodical function $f(x)$ at all places instantly.
2. The QFT is computed by a quantum circuit which uses:
   • Hadamard transform [31], a square matrix consisting of +1 and −1, and the rows are orthogonal to each other. The Hadamard transformation is

- at most times a sort of 'pre-processing step' in most of the quantum algorithms; it maps n qubits (0 or 1) to a superposition of 2^n orthogonal states.
- Quantum gates [32] (time-invertible quantum circuits operating on a set of qubits).

The quantum computer can be installed in different states and in certain ways contribute to the value of 'r' for a certain 'possible' value of the period 'r'. These countries cancel each other ultimately. However, the states add up in the same direction only for the right value of 'r'. The algorithm of Shor is of probabilistic character and enhances its performance with repeats. In order to deliver the findings of the main factorizing issue in only few seconds, Shor's method skilfully leverages the effects of quantic parallelism, although a classical computer would take more than an eternity for the observed world in certain situations to do this. This is the main reason to create the quantum cryptography. Examine Table 9.2 for the impact of large-scale quantum computer [33] on the traditional cryptographic algorithm.

It is crystal clear that the traditional cryptographic algorithm as illustrated in Table 9.2 will not stand against the quantum computer. In order to resist the quantum computers, new cryptographic algorithms need to be designed and the fundamentals of discrete logarithms should not be addressed [34] or the major difficulty of disintegration like the established cryptographic algorithm, such cryptosystem, will use the quantum properties to act as novel cryptography as quantum cryptography. If one needs to break such a cryptosystem, he/she has to break the laws of quantum physics [35]. As a matter of fact, if one successfully breaks the laws of quantum physics, they will be called by Swedish Academy and asked to receive their Nobel Prize for Physics; it is seriously tough to break the laws of quantum physics. To implement proper information system security, we must include a framework of fundamental systematic processes such as cryptography and network security. Using the help of quantum physics as foundation and adding the pillars of cryptographic principles, researchers caused a major development in the cryptographic field, and it is referred to as quantum cryptography. The safety of information exchange can be assured by two theories of quantum physics. One is Heisenberg's uncertainty principle, and the other is quantum no-cloning theory. The foremost objective of the research of

TABLE 9.2
Cryptographic Algorithms

Cryptographic Algorithm	Type	Purpose	Impact of Large-Scale Quantum Computer
AES	Symmetric key	Encryption	Larger key sizes needed
SHA-2 and SHA-3	Not applicable	Hash functions	Larger output needed
RSA	Public key	Signatures and key establishment	No longer secure
ECDSA and ECDH (elliptic-curve cryptography)			
DSA (finite field cryptography)			

quantum cryptography is to outline the cryptographic algorithms and respective protocols, mostly targeted by the novel quantum computing attacks.

In the year 1969, Wiesner coined the idea of quantum money, which is now acting as the backbone for quantum cryptography. Restricted by the technology in that time, even though the idea of quantum money was very creative, it could not be materialized, which made it remain unpublished till the year 1983. The initial applied quantum key distribution (QKD) protocol [36] was suggested by Bennett and Brassard, in the year 2011. Playing on the support of single-photon polarization, the researchers founded the implementation of the protocols in the QKD after a Himalayan struggle was added up in the QKD to gear towards advancing the security and its proficiency. In the year 1991, Artur Ekert, British professor of quantum physics, recommended a set of rules; later in the year 1992, Bennett considered the idea of deploying any two nonorthogonal states. The improvement is highly effective and simpler. Succeeding this, several QKD protocols employing the foundation of quantum physics have been formulated in succession. The oblivious transfer protocol, as a significant cryptographic basic protocol, is one of the main cryptographic privacy methods. The protocol that provides in which the source sends a vast volume of possible information to the recipient, but is unaware of the precise subject of the communication. Cr'epeau originally put out the notion of quantum overlooked transfer (QOT) [37] in 1994.

A useful modification was made to the quantum oblivious transfer protocol in 1994, by which they were able to validate that quantum mechanics security over individual measurements works regardless of whether or not one is aware of being observed. The QOT protocol was shown to be secure in the presence of an adversary in 1998, when the protocol was presented. There were several methods put forth for increasing the QOT effectiveness, some of which were more effective than others. It's also a member of the class of quantum cryptography protocols, known as QA protocols [38]. The proposal to expand was suggested in 2001. Since then, a number of QA processes have also been proposed one after the other. The quantum cryptography protocol has branched out into several paths presently. Aside from the protocols (i.e. QKD protocol, QOT protocol, and QA protocol) that we have already described, other quantum cryptography protocols comprise quantum bit commitment (QBC) protocols [39] and quantum signature (QS) protocols [40].

While quantum interactivity and data handling have their own advantages, they are often bigger in size than the conventional ones, which is founded in the specifics of quantum systems. The properties of quantum information are essentially constituted by the following:

i. **Uncertainty principle:** German scientist Werner Heisenberg (1927) was the originator of the uncertainty principle. The essential premise of the uncertainty principle [41] is that it is difficult to predict where a particle is located in the microcosm, and the particle is thus assumed to be present in several locations at once with a given probability.

ii. **Quantum no-cloning theory**: This undefined quantum state's inability to be replicated and/or deleted is what makes it so useful. Cloning is a process that takes an undistinguishable quantum state in one system and duplicates it in another. In previous research, it was discovered that engines capable of replicating quantum systems do not exist. A universal principle of undeleting pledges that even if the adversary erases or destroys quantum states on the information medium, no further signal of it remains to leak into secure communication. It was advised that one may not repeat a random quantum state, which is clearly unallowable according to the principles of linearity in quantum theory.

iii. **Quantum teleportation:** The responder measures the quantum condition of the original data [42], and the source is stated in the manner of traditional communication, which provides conventional intelligence. Quantum data are the remainder of the data not obtained by the source in the measure and are transferred by evaluation to the recipient. In 1993, the system is advocated for the television of an uncertain quantum state.

iv. **Hidden characteristics of quantum information:** This has remarkable qualities which are not maintained by ordinary data. In particular, the local measuring operation, which can only be indicated by joint observation, cannot obtain the quantum code data in the entangled state. Works have been suggested on the disguise of quantum states.

The quantum interaction is divided into two components.

- Quantum direct communication
- Quantum teleportation communication.

The easiest way for the transmission of quantum signals is the direct transmission model [43], as shown in Figure 9.6.

Let's look at the graphic depiction above. There, we can see that Alice must connect with Bob over a quantum channel. Alice initially had to create a series of photons in the quantum direct transmission paradigm through the preparatory device as per the data that Bob wants to convey. This knowledge should also be processed by the quantum source encoder and the encoder for quantum corrections to errors after the production of the quantum source (QECC). The quantum information can then be immediately sent to the quantum channel (optical fibre). Here, outside noise readily

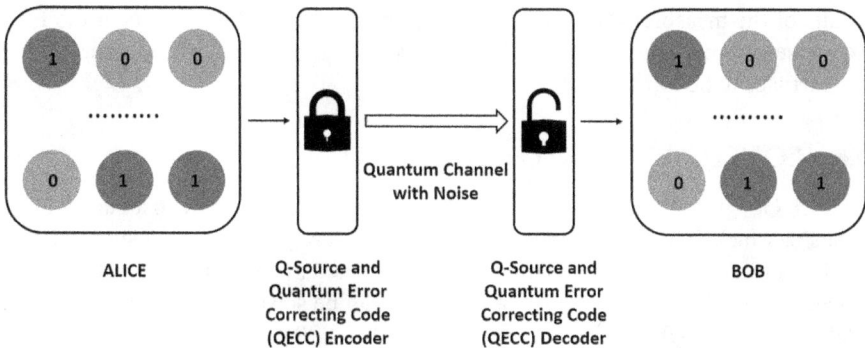

| ALICE | Q-Source and Quantum Error Correcting Code (QECC) Encoder | Q-Source and Quantum Error Correcting Code (QECC) Decoder | BOB |

FIGURE 9.6   Representation of quantum direct communication model.

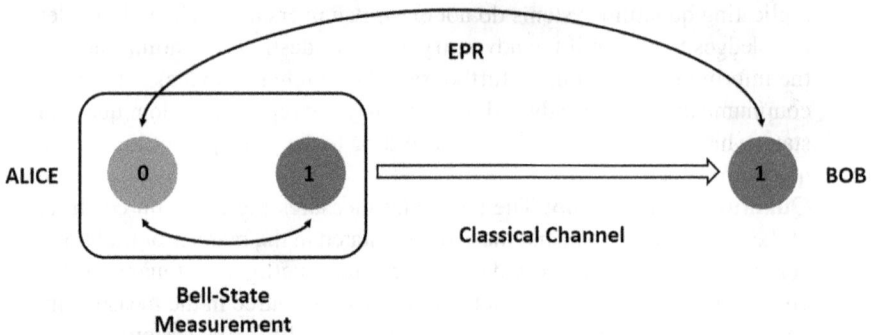

FIGURE 9.7 Representation of quantum teleportation.

disturbs the quantum channel. Therefore, the Bob receiver encodes the QECC to the received signal at first and then encodes the quantum source. Finally, the initial quantum message is received from Bob.

When we're talking about the two methods of quantum communication, we have the teleportation method as illustrated in Figure 9.7. Instead of being based on classical communication, quantum computing uses qubits that are capable of being in a range of orthogonal superposition states, and these qubits are also capable of being in the entangled state. The main idea behind quantum teleportation is to build a quantum channel by employing the maximal entangled state of two particles. This occurs when the message is conveyed via a quantum operation [44]. It is important to remember that the different communication channels used (teleportation vs. direct conversation) are the only difference between the two options. A visual illustration of the quantum teleportation model is shown in Figure 9.7.

In this model, we show Alice, who desires to communicate one-bit quantum with Bob in another location, as she appears in the following diagram. The first step is the production of an EPR pair from the EPR entanglement source. The next step in the process is for one of the particles to be delivered to Alice, and the other is conveyed to Bob over the quantum channel. After then, in order to send information, Alice needs to gauge the particles in the EPR entangled pairs [45] and the pending bits she possesses. Alice next goes on to tell Bob of measuring results. Furthermore, the results of the measurement that was conducted by Alice and the results of the EPR experiment agree. This means that Bob will be able to retrieve the information about the particles to be transported.

9.6 CONCLUSIONS AND FUTURE OF QUANTUM

In 2019, Google claimed that they achieved quantum supremacy with an array of 54 qubits of which 53 were functional. They tested with sequences of problems and calculations using the quantum computer and completed the operations in <4 minutes. Google claimed that the same set of problems can be solved by a supercomputer in 10,000 years. But IBM reported that the claim of 10,000 years for a supercomputer is a hugely exaggerated figure and they added that the supercomputer will take 60 hours to complete the same set of problems. Still, the 4 minutes completion by quantum computing compared with 60 hours of supercomputing is considerably promising. By the end of 2020, a group of researchers from University of Science and Technology of China (USTC) accomplished quantum supremacy by employing a type of boson sampling on 76 photons with their photonic quantum computer, Jiuzhang, which has magnificently conducted a Gaussian boson sampling task in 200 seconds, whereas the same task is expected to be completed in an estimated time of half a billion years on the fastest traditional non-quantum computer. The USTC researchers claimed that Jiuzhang is 10 billion times faster than Google's superconductor-based Sycamore processor, becoming the second computer to attain quantum supremacy.

Researchers strongly suggested that quantum computers will be used by huge organizations only and not by normal people. I want to remind you that the same thing was said by researchers in the early 1960s to 1980s that the digital computers or traditional computers will be used by research organizations and military. But what we witness now is completely different. Hopefully and optimistically, we need to wait for a few more years to witness the real-life operations of quantum computers and the usage of quantum cryptography.

REFERENCES

1. R. J. Hughes, D. M. Alde, P. Dyer, G. G. Luther, G. L. Morgan, and M. Schauer, "Quantum cryptography," *Contemp. Phys.*, vol. 36, no. 3, pp. 149–163, 1995, doi: 10.1080/00107519508222149.
2. L. Jian, L. Na, Z. Yu, W. Shuang, D. Wei, C. Wei and M. Wenping., "A survey on quantum cryptography," *Chinese J. Electron.*, vol. 27, no. 2, pp. 223–228, 2018, doi: 10.1049/cje.2018.01.017.
3. J. Preskill, "Quantum computing: Pro and con," *Proc. R. Soc. A Math. Phys. Eng. Sci.*, vol. 454, no. 1969, pp. 469–486, 1998, doi: 10.1098/rspa.1998.0171.
4. S. Arunachalam and R. De Wolf, "A survey of quantum learning theory," arXiv, January 24, 2017.
5. S. A. Selvi, R. S. Rajesh, and M. A. T. Ajisha. "An efficient communication scheme for Wi-Li-Fi network framework." *In 2019 Third International Conference on I-SMAC (IoT in Social, Mobile, Analytics and Cloud)(I-SMAC)*, pp. 697–701, Palladam, India, IEEE, 2019.
6. C. Baily and N. D. Finkelstein, "Development of quantum perspectives in modern physics," *Phys. Rev. Spec. Top. Phys. Educ. Res.*, vol. 5, no. 1, p. 010106, 2009, doi: 10.1103/PhysRevSTPER.5.010106.
7. S. B. McKagan, K. K. Perkins, and C. E. Wieman, "Design and validation of the quantum mechanics conceptual survey," *Phys. Rev. Spec. Top. Phys. Educ. Rese.*, vol. 6, no. 2, p. 020121, 2010, doi: 10.1103/PhysRevSTPER.6.020121.
8. C. Baily and N. D. Finkelstein, "Refined characterization of student perspectives on quantum physics," *Phys. Rev. Spec. Top. Phys. Educ. Res.*, vol. 6, no. 2, p. 020113, 2010, doi: 10.1103/PhysRevSTPER.6.020113.
9. N. Zettili and I. Zahed, "Quantum mechanics: Concepts and applications," *Am. J. Phys.*, vol. 71, no. 1, pp. 93–93, 2003, doi: 10.1119/1.1522702.
10. T. D. Ladd, F. Jelezko, R. Laflamme, Y. Nakamura, C. Monroe, and J. L. O'Brien, "Quantum computers," *Nature*, vol. 464, no. 7285, pp. 45–53, 2010, doi: 10.1038/nature08812.
11. R. Raussendorf and H. J. Briegel, "A one-way quantum computer," *Phys. Rev. Lett.*, vol. 86, no. 22, pp. 5188–5191, 2001, doi: 10.1103/PhysRevLett.86.5188.

12. L. Gyongyosi and S. Imre, "A survey on quantum computing technology," *Comput. Sci. Rev.*, vol. 31, pp. 51–71, 2019, doi: 10.1016/j.cosrev.2018.11.002.
13. G. Carrascal, A. A. del Barrio, and G. Botella, "First experiences of teaching quantum computing," *J. Supercomput.*, vol. 77, no. 3, pp. 2770–2799, 2020, doi: 10.1007/s11227-020-03376-x.
14. S. Kak, "A three-stage quantum cryptography protocol," *Found. Phys. Lett.*, vol. 19, no. 3, pp. 293–296, 2006, doi: 10.1007/s10702-006-0520-9.
15. E. Knill, R. Laflamme, R. Martinez, and C. Negrevergne, "Benchmarking quantum computers: The five-qubit error correcting code," *Phys. Rev. Lett.*, vol. 86, no. 25, pp. 5811–5814, 2001, doi: 10.1103/PhysRevLett.86.5811.
16. L. K. Grover, "Synthesis of quantum superpositions by quantum computation," *Phys. Rev. Lett.*, vol. 85, no. 6, pp. 1334–1337, 2000, doi: 10.1103/PhysRevLett.85.1334.
17. T. G. Noh, "Counterfactual quantum cryptography," *Phys. Rev. Lett.*, vol. 103, no. 23, p. 230501, 2009, doi: 10.1103/PhysRevLett.103.230501.
18. J. Preskill, "Quantum computing and the entanglement frontier," March 2012, Accessed: February 22, 2021. [Online] Available: http://arxiv.org/abs/1203.5813.
19. Č. Brukner and A. Zeilinger, *Quantum Physics as a Science of Information*. Springer: Berlin, Heidelberg, 2005, pp. 47–61.
20. L. O. Mailloux, C. D. Lewis, C. Riggs, and M. R. Grimaila, "Post-quantum cryptography: What advancements in quantum computing mean for IT professionals," *IT Prof.*, vol. 18, no. 5, pp. 42–47, 2016, doi: 10.1109/MITP.2016.77.
21. A. Vaziri, G. Weihs, and A. Zeilinger, "Superpositions of the orbital angular momentum for applications in quantum experiments," *J. Opt. B Quantum Semiclassical Opt.*, vol. 4, no. 2, p. S47, 2002, doi: 10.1088/1464-4266/4/2/367.
22. L. DiCarlo, J. M. Chow, J. M. Gambetta, L. S. Bishop, B. R. Johnson, D. I. Schuster, J. Majer, A. Blais, L. Frunzio, S. M. Girvin and R. J. Schoelkopf, "Demonstration of two-qubit algorithms with a superconducting quantum processor," *Nature*, vol. 460, no. 7252, pp. 240–244, 2009, doi: 10.1038/nature08121.
23. A. C. C. Say and A. Yakaryilmaz, "Magic coins are useful for small-space quantum machines," 2014, Accessed: February 22, 2021. [Online] Available: http://arxiv.org/abs/1411.7647.
24. E. W. Piotrowski and J. Sladkowski, "The next stage: quantum game theory," 2003, Accessed: February 23, 2021. [Online] Available: http://arxiv.org/abs/quant-ph/0308027.
25. A. Nanda, D. Puthal, S. P. Mohanty, and U. Choppali, "A computing perspective of quantum cryptography energy and security," *IEEE Consum. Electron. Mag.*, vol. 7, no. 6, pp. 57–59, 2018, doi: 10.1109/MCE.2018.2851741.
26. V. Vedral, "Quantum entanglement," *Nat. Phys.*, vol. 10, no. 4, pp. 256–258, 2014, doi: 10.1038/nphys2904.
27. C. Elliott, "Quantum cryptography," *IEEE Secur. Privacy*, vol. 2, no. 4, pp. 57–61, 2004, doi: 10.1109/MSP.2004.54.
28. J. H. Cheon, D. Kim, J. Lee, and Y. Song, "Lizard: Cut off the tail! A practical post-quantum public-key encryption from LWE and LWR," in Lecture Notes in Computer Science (including subseries Lecture Notes in Artificial Intelligence and Lecture Notes in Bioinformatics), September 2018, vol. 11035 LNCS, pp. 160–177, doi: 10.1007/978-3-319-98113-0_9.
29. T. Monz, D. Nigg, A. Esteban, F. Brandl, S. Philipp, R. Richard, X. Shannon, L. Isaac and B. Rainer, "Realization of a scalable Shor algorithm," *Science*, vol. 351, no. 6277, pp. 1068–1070, 2016, doi: 10.1126/science.aad9480.
30. D. Camps, R. Van Beeumen, and C. Yang, "Quantum Fourier transform revisited," *Numer. Linear Algebr. Appl.*, vol. 28, no. 1, p. e2331, 2021, doi: 10.1002/nla.2331.
31. E. Biham, G. Brassard, D. Kenigsberg, and T. Mor, "Quantum computing without entanglement," *Theor. Comput. Sci.*, vol. 320, no. 1, pp. 15–33, 2004, doi: 10.1016/j.tcs.2004.03.041.

32. J. Preskill, "Quantum computing in the NISQ era and beyond," *Quantum*, vol. 2, p. 79, 2018, doi: 10.22331/q-2018-08-06-79.

33. A. G. Fowler, M. Mariantoni, J. M. Martinis, and A. N. Cleland, "Surface codes: Towards practical large-scale quantum computation," *Phys. Rev. A Mol. Opt. Phys.*, vol. 86, no. 3, p. 032324, 2012, doi: 10.1103/PhysRevA.86.032324.

34. X. Wen, X. Niu, L. Ji, and Y. Tian, "A weak blind signature scheme based on quantum cryptography," *Opt. Commun.*, vol. 282, no. 4, pp. 666–669, 2009, doi: 10.1016/j.optcom.2008.10.025.

35. V. Scarani and C. Kurtsiefer, "The black paper of quantum cryptography: Real implementation problems," *Theor. Comput. Sci.*, vol. 560, no. P1, pp. 27–32, 2014, doi: 10.1016/j.tcs.2014.09.015.

36. R. Valivarthi, I. Lucio-Martinez, P. Chan, A. Rubenok, C. John, D. Korchinski, C. Duffin, F. Marsili, V. Verma, MD. Shaw, JA. Stern, SW. Nam, D. Oblak, Q. Zhou, JA. Slater and W. Tittel., "Measurement-device-independent quantum key distribution: From idea towards application," *J. Mod. Opt.*, vol. 62, no. 14, pp. 1141–1150, 2015, doi: 10.1080/09500340.2015.1021725.

37. Y. Song and L. Yang, "Practical quantum bit commitment protocol based on quantum oblivious transfer," *Appl. Sci.*, vol. 8, no. 10, p. 1990, 2018, doi: 10.3390/app8101990.

38. T. S. Wei, C. W. Tsai, and T. Hwang, "Comment on 'quantum key distribution and quantum authentication based on entangled state,'" *Int. J. Theor. Phys.*, vol. 50, no. 9, pp. 2703–2707, 2011, doi: 10.1007/s10773-011-0768-0.

39. X. Sun, F. He, and Q. Wang, "Impossibility of quantum bit commitment, a categorical perspective," *Axioms*, vol. 9, no. 1, p. 28, 2020, doi: 10.3390/axioms9010028.

40. J. M. Arrazola, P. Wallden, and E. Andersson, "Multiparty quantum signature schemes," *Quantum Inf. Comput.*, vol. 16, no. 5–6, pp. 435–464, 2015, Accessed: Februay 23, 2021. [Online] Available: http://arxiv.org/abs/1505.07509.

41. M. S. Sharbaf, "Quantum cryptography: A new generation of information technology security system," in *ITNG 2009-6th International Conference on Information Technology: New Generations*, 2009, pp. 1644–1648, doi: 10.1109/ITNG.2009.173.

42. C. Y. Chen, G. J. Zeng, F. J. Lin, Y. H. Chou, and H. C. Chao, "Quantum cryptography and its applications over the internet," *IEEE Netw.*, vol. 29, no. 5, pp. 64–69, 2015, doi: 10.1109/MNET.2015.7293307.

43. X. B. Chen, T. Y. Wang, J. Z. Du, Q. Y. Wen, and F. C. Zhu, "Controlled quantum secure direct communication with quantum encryption," *Int. J. Quantum Inf.*, vol. 6, no. 3, pp. 543–551, 2008, doi: 10.1142/S0219749908003566.

44. D. Collins, N. Gisin, and H. De Riedmatten, "Quantum relays for long distance quantum cryptography," *J. Mod. Opt.*, vol. 52, no. 5, pp. 735–753, 2005, doi: 10.1080/09500340412331283633.

45. J. Li, N. Li, L. L. Li, and T. Wang, "One step quantum key distribution based on EPR entanglement," *Sci. Rep.*, vol. 6, p. 28767, 2016, doi: 10.1038/srep28767.

10 An Extensive Exploration of Privacy and Compliance Considerations for Hybrid Cloud Deployments

P. Hemalatha
IFET College of Engineering

W.T. Chembian
Gojan School of Business and Technology

A. Ranjith Kumar
Lovely Professional University

S. Ramesh
Krishnasamy College of Engineering & Technology

B. Jegajothi
Sri Venkateswara College of Engineering

CONTENTS

DOI: 10.1201/9781003219880-10

The evolution of hybrid cloud environments in the past few decades has made it tough for visibility, interoperability and consistent controls because of the demanding considerations. At the same time, however, cybercrime rates have rapidly increased, breaching the cloud storage and having access to personal data. Current research works focus on centralized and decentralized access control methods in order to safeguard the confidential data. The studies mainly aim at improving the integrity, security and privacy of the data. Some examples are application of fuzzy-based methods in cloud to strengthen the privacy by using key word searches to assess the information theft. Additionally, the use of homomorphic encryption methods produces ciphertext from getting compromised. However, there have been challenges while transferring data across automated and self-provisioned settings. The recent and the most commonly used methods to ensure security are auditing, data and network security and mobility, integrating identity and access management sources.

Similarly, compliance requirements are met by applying automated policy and centralized framework. Certain risk mitigation measures taken in data centers are virtualization. Virtualization is the basic strategy to allow certain devices to access the hypervisor admin account to minimize the chances of exploitation. In order to carry out scalability, fulfilling the demands of IT field, imposing security rules combines to form segmentation tool and task definition. Segmentation is vital for allocating policy settings and fulfills the operational needs that work in the execution setting, and it is essential for assessing the risk of hybrid cloud environments. Some of the hybrid cloud security challenges are as follows:

Data protection: During the data transmission limit data exposure found very challenging. The data are replicated in more instances during encryption. Hence, strong security policies have been enforced to overcome the disadvantages.

Compliance and governing the data: In deployments such as healthcare and government sectors, the hybrid cloud needs additional considerations. It is highly recommended to check the decentralized environments for their compliance. And also, it is mandatory to streamline the security baselines for preparing security audits.

10.1   PRIVACY AND SECURITY ISSUES IN HYBRID CLOUD ENVIRONMENTS

During these decades, the globe has vastly turned into a digital space. As the amount of data keeps expanding each day, researchers are challenged to develop higher versions of data storage tools and techniques. Cloud computing is capable of driving and transforming companies through business intelligence, predict inventory requirements, effective cost ownership and target marketing plans. In cases of high-performance data computing (HPDC) setting, transfer of data, processing data and storing it in the cloud are unavoidable to assess the hidden patterns. Nevertheless, according to regulatory policies, these applications come along with certain confidentiality problems in a hybrid cloud environment, while dealing with private datasets. For example, (i) the process of producing transparency to individuals leads to disclosing the algorithmic methods by passing on the necessary information and analytical algorithm. (ii) Carrying out an informed consent forms a lawful responsibility by providing necessary information to the user. (iii) Deleting and restoring of data according to user request causes leak of data in the data traffic. (iv) Re-identification and scanning huge datasets leads to identity theft. In addition, certain cloud storage systems are now known to be primary attack targets because of the increased number of cyberhackers.

Ensuring a cybersecurity system in intensive applications still poses to be one of the major challenges due to the rising number of data breaches, and cyberattacks are multifaceted. Despite having potentially strong security tools and practices, cybercrimes still prevail in the hybrid cloud environment. Distributed frameworks are open-sourced and are associated with lesser security policies. These are used by large data users and are devoid of any protection. Furthermore, the use of NoSQL databases and endpoints is not secure and valid; therefore, authenticity is not considered.

In addition, one other chief cybersecurity issue is data mining solutions, for pattern predictions which enable insiders to have access, thus increasing the scope of abuse. Successively, the use of real-time security systems and compliance tools that contain a huge amount of data may make it easier for the cyberhackers to invade sensitive assets through data breaches. Besides that, some of the other existing issues in the present hybrid cloud applications are inefficiency in data protection caused by lack of advanced tools, several interventions used in cluster networks and inefficient monitoring techniques to rule out suspicious malware. Today, there stands a need to protect the HPDC setting from privacy and cybersecurity threats faced in hybrid cloud environment.

10.2 IDENTITY MANAGEMENT (IM) IN HYBRID CLOUD SETTINGS

Identity management (IM) is effectively used to carry out the following functions:
- Data administration
- Data discovery in cloud
- Documentation
- Policy enforcement
- Data management
- Data transfer and authentication.

Contrastingly, the application of identity and access management (IAM) ensures that a particular identity is installed in all functions to promote security in the same time. This method helps in identifying the users and devices and provides access to the data and resources. In cases where the applications have access to the hybrid, device or service exhibiting identity is not a requirement anymore to validate itself. Instead, individual identity verification procedure can be obtained by an identity provider that could help in reducing the load of the application. Large-scale distribution systems can be simplified by using identity and access management. This is mostly carried out in organizations and business correspondences. Certain private companies use this between their organization and cloud [1].

IAM consists of a huge organizational setup that is enabled to identify cloud objects based on prior policies [2]. Identity and access management is carried out by several functions. These comprise provisioning, managing authenticity, managing federated identity and managing compliance and authorization [3]. Operational areas ensure that the authorized users are securely incorporated into the cloud environment. An XML-based framework also called Service Provisioning Markup Language (SPML) is successfully used for the process of identity management. The operations performed by it are transfer of resources, users and service information between companies. One drawback of SPML is the application of multi-proprietary protocols from several vendors toward a common application programming interface (API). Interaction among APIs is not possible because they belong to different vendors. The origin of different APIs makes sure that log-in data such as passwords and digital certificates are safe [4].

The next operation of interest is federated identity management that helps cloud with an organization's identity provider. It functions by transferring public key infrastructure (PKI), thus ensuring privacy, integrity and assurance between identity provider and web-based application. The final decision regarding granting access to the user depends upon the authorization management after the successful authentication process takes place. The last step is known as compliance management, which ensures the privacy of the organization's data and functions based on the existing policies and regulations [5]. Privacy and interoperability are the two major issues pertaining to the current identity management methods, especially in the public cloud environment [6]. At present, IAM systems are well equipped to minimize threats related to the cloud storage. Hence, IAM is largely used in many organizations to protect their vital data by having a control over every individual's access permission [7].

Certain frequently used IAM providers are SailPoint, Oracle, RSA, IBM and Core Security. SailPoint is capable of functioning with managing passwords, controlling compliance, data access governance, requesting access, automated provisioning and single sign-on [8]. IBM Identity provides web access request solutions, provisioning users, authenticating multiple factors, enterprise single sign-on, identity and access control, user activity and compliance [9]. The IBM provides with request solutions, provisioning users, authenticating multiple factors, enterprise single sign-on, identity and access control and compliance of users and activities [9]. However, Oracle IAM provides four solutions related to cloud safety. The first solution products carry out identity administration such as self-service account request, managing identity life cycle, self-service account, and password and enterprise role. Oracle IAM system offers solutions such as privacy, identity federation and single sign-on. With respect to access control, a third solution such as the fine-grained entitlements, risk-based authorization and web security is provided. A fourth solution is also proposed by Oracle IAM that recognizes and assesses governance that includes duty delegation, audits, reporting compliance, conflict resolutions, attestation, identifying and preventing frauds and directory services (identity virtualization, great storage, database security and synchronization) [10]. RSA SecurID Suite is capable of offering authenticity access management and governance identification, life cycle management and analysis of risks. Security plays a major role in functioning with identity management and accessing governance in compliance, privileged services, managing passwords, controlling access and managing identity. Privileged identity management (PIM) is responsible for managing the super-user accounts and account holder rights. PIM is equipped with provisional tools and special PIM products for the purpose of identity management. A few well-known PIM providers are IBM PIM, CyberArk and Oracle PAM.

10.3 SAFEGUARDING DATA TRANSFER AND WORKLOADS ACROSS HYBRID CLOUD SETTINGS

The exchange of data from the data center to cloud comes with many issues. One of the major problems is that the idea of having a migratory security control is developed in the later stages, by the end of the procedure rather than right from the beginning. According to the Amazon Web Services (AWS), data center managers are highly accountable for all the data processed in AWS to be kept confidential. There are also certain misperceptions regarding locking the infrastructure that causes securing of workloads. Therefore, it is essential for companies to develop security controls and ensure protection of public data from threats.

Some of the data exfiltration attacks are the following:
- AWS attacks on public subsets and public cloud.
- Attack on open ports, thus gaining access to user's Internet-facing assets.
- An attempt to obtain private servers by lateral movements.
- Downloading data and functioning servers in AWS.
- AWS environment has enabled controls that provide security from malicious activities; however, all of these activities cannot be prevented. These controls lack cloud visibility and on-premises settings.

10.3.1 MIGRATION OF THREATS TO THE CLOUD

Invaders are aware of the situations pertaining to transfer of data from organizations to the cloud, its security issues and workload through the AWS setting. It is also known to them that the data center managers are not very efficient toward providing data security. The Internet-facing workloads are not placed in the cloud unlike the high protection that is mostly used in the premises of data centers. Following are certain frequently made errors by organizations while transferring workloads to cloud:

- **False perception regarding security being provided by the provider**: The AWS platform has specified information regarding the areas which will be provided by their security, the rest being the responsibility of the customers. Nevertheless, not many organizations are aware of this and the IaaS shared responsibility. The organizations have to share the responsibility, especially for the application of software to be installed above infrastructure.

- **Assuming cloud process to be similar to on-premises**: The algorithm of workflow varies largely for cloud and on-premises environment. However, organizations assume that functioning of cloud and premise is the same.
- **Improper transfer of security controls with workloads**: It becomes impossible to enable defense in depth as organizations transfer these workloads based on on-premises environment. The agenda behind setting up multiple security controls is to maximize the ability to detect threats.
- **Transferring inconsistent security policies**: An example of one such method of transferring improper security policy is 'lift and shift.' Many organizations practice this method in order to shift data to cloud. According to this, the workload is shifted 'as-is,' leading to non-formation of a new workload. Even though this method is faster, it causes workload exposure.

Below-mentioned are some situations that illustrate the workload-related attacks in public cloud domain:

1. Attacker takes advantage of unsafe web server situated in AWS, hence obtaining a shell access.
2. Internal network scanning is carried out by bad actor in the web server, which indicates the presence of phpMyAdmin in a database.
3. Attacker gains entry into the database server by having access to default/ weak passwords in phpMyAdmin (an open-source admin tool), resulting in the attacker to have a complete approach toward the database contents.
4. After this, the database contents are extracted and its copies are made in web server, making it ready for exfiltration.
5. Ultimately, the copied data are uploaded to an external sever and the attacker earns revenue from targets data either by trading it to a competitor company or by initiating secondary attacks against target's customers.

Attacks that occur at various stages as described above lead to security threats for organizations. A huge amount of investment is made for securing the on-premises data and assets. However, this security does not assure for cloud workloads. Hence, organizations are open to the malicious threats and attacks.

10.4 STORING DATA IN A THIRD-PARTY OR HOSTED ENVIRONMENT

The most significant asset to a company is the data, and therefore lots of efforts are directed toward ensuring its security. To protect data from unauthorized users, access control is practiced [11]. Additionally, other programs such as data leakage and privacy methods must be implemented. Cloud service providers must also be refrained from having access to organization's data, which needs higher safety measures [12]. An efficient solution for these limitations would be encryption strategies to promote privacy and confidentiality of data. Despite all this, key management in encryption still remains an unresolved issue [13,14]. An effective solution may comprise effective encryption techniques. However, key management plays a vital role in the encryption process [13,14]. Cryptographic keys should be securely stored.

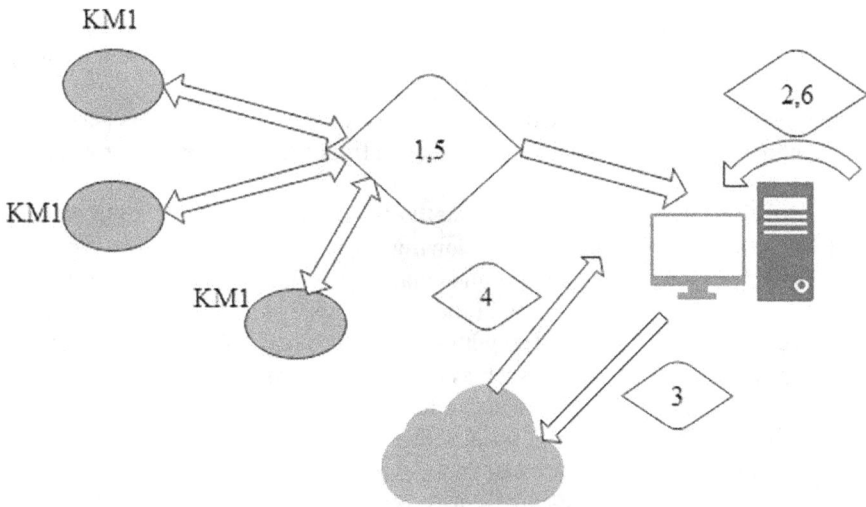

FIGURE 10.1   Shamir's threshold scheme.

Any kind of mismanagement will cause data loss. Hence, cryptographic keys should be securely placed in a standardized manner [10]. One of the very well-known and successful data security schemes for managing cryptographic key servers is Shamir's (k, n) threshold scheme [19]. This makes use of the k and n shares to rebuild the key. Documentation of policies is essential that should contain information about the key and data. Random symmetric keys are made to work on encryption and integrity. It is the role of the key manager (KM) to keep the symmetric keys safe (Figure 10.1), which is then deleted by the user. Once the data are encrypted and the key is secured, it is then uploaded to the cloud. In case the data and keys have to be downloaded, the user will have to provide a policy file. This step is succeeded by decrypting the key by a KM. The steps can be summarized as follows.

1. The KM receives the public keys.
2. The role of client is to break up the symmetric key S into n shares $(S_1, S_2, ..., S_n)$, followed by encrypting the ith share with the public key of ith KM – delete S.
3. Upload all S shares to the cloud.
4. The user may now download all the key shares from the cloud.
5. The role of the user is to randomly pick up 'k' numbers of KMs – followed by sending ith share of S to ith KM. After this, the user will receive back the decrypted ith share.
6. Using Shamir's method to reconfigure S from k shares.

In [11], the authors presented an algorithm to safeguard the cloud data. It provides the users with data integrity and availability. A gateway application was used to manage the data integrity. Iris – a file system – was developed to move the organizational files to cloud. In addition to that, another gateway known as Merkle tree provides data with freshness and integrity by combining with file blocks, MAC and file versions at various levels. Merkle tree also operates with cryptographic keys and confidentiality. Also, an audit framework described in [11] indicates and assures data freshness irreversibility and resistance toward disk failures. However, this proposal is dependent on user confidentiality and cannot assure protection from wholesale service providers.

Application of the above-stated methods will help us store data in a third-party environment in the following ways.

i. Developing and stating data confidentiality principles that meet the level of sensitivity and confidentiality of the stored data. For this purpose, stronger principles can be used to deal with sensitive data and safeguard its identity. Administrative safeguards such as evaluation of risks and its management and restricting access, technical safeguards such as data encryption and authenticating users and physical safeguards such as securing devices are used for protecting confidential data.

ii. Implementing these safeguards to protect data from getting deleted by unauthorized users, modifications or fabrication. This plays a vital role as devices in cloud can be interoperated. Permissions are minimized to control the access of users by identifying them and providing them with appropriate data access.

iii. Develop and maintain organizational data security policies. Data are said to be safe if they have an efficient administration with physical and technical security. These are essential requirements while functioning in a cloud or a server setting. However, in cloud setting, the resources carrying out server maintenance and backup can also be used for supporting security procedures.

iv. Study and note down the network risk analysis and develop a risk management strategy. Duties such as security and risk management must be delegated to professionals who should undergo orientation and training about data users. They must also make sure that the employees of the organization understand and practice the contingency plans for security emergencies. Government officials may have access to the data via legal notices for legal aid or data requests.

v. Monitoring the implementation process and carrying out compliance audits. The purpose of this is to evaluate security vulnerabilities and also to develop and execute policies to guide workforce in accordance with security breaches. The organizations can also provide their employees with training on data handling.

vi. Practicing efficient practices toward handling sensitive and private information to be stored in cloud. Practicing the guidelines issued by national and international regulations regarding handling and storing data. Assessing the policies that allow data storage in cloud.

10.5 SECURITY THREATS AND COMPLIANCE MODELS

Several threats related to security issues are stated in [16] and [17] and were taken as references to study the threats mostly faced by cloud users. The findings that were received are matched with security controls and compliance.

Cloud security threats mainly consist of the following:

1. **Breaching of data**: This has impact on the data privacy and organization. Data should be safe from unauthorized users.
2. **Loss of data**: This may take place due to failure of hardware or attacks in the system. To intervene in such scenarios, data backup must be applied.
3. **Hijacking accounts/service traffics**: The user's integrity is altered. Private information of the user may be at risk.
4. **Insecure interfaces and APIs**: Interfaces and APIs are used for communication between users and providers. APIs must be capable of encrypting and transferring data through interfaces.
5. **Denial of service**: Interruption by an unauthorized user can cause alteration in the encryption key, thus lowering the system speed. In order to mitigate such attacks, it's the responsibility of both the user and the cloud to develop an algorithm to prevent such malicious activities.
6. **Malicious insider personnel**: These are employees working within the organization. Depending upon the organizational policies, legal action against such employees may be taken.
7. **Misusing cloud services**: Misusing cloud features such as multi-tenancy to have an access to data of other organizations. Measures can be taken by cloud to prevent such activities.
8. **Lack of diligence**: Organizations all over the world prefer using cloud services; however, they are not completely aware of its threats. Hence, awareness programs must be conducted to educate all the fields of cloud services.
9. **Sharing technological vulnerabilities**: Cloud shares their resources in a scalable manner which is ideal and hence should be practiced in all domains including monitoring systems.

The author in [18] presented various privacy and security problems that have adverse effects on cloud computing. This paper takes into consideration issues such as governance, identity, architecture, compliance, access control, data protection, availability and incident reporting. Hence, it can be inferred that cloud computing compliance is the major problem that should be mitigated with policies, which may vary from one country to the other. The author [18] stated that it is challenging to impose regulations in cloud. However, the effect of data location was evaluated along with control loss and public transparency. The other limitations mentioned are collection, identification, processing and analysis, and protection of the data stored. Nevertheless, the paper does not describe mapping techniques used for complex policies to the best practices and patterns/RAs. Cloud ensures compliance by obtaining a certification from a third party. According to the present study, it can be said that third-party auditors make use of proprietary solutions. These are devoid of vendor-neutral models that are used as checklists. The authors in [19] made a comparison between GLBA, HIPPA, PCI and SOX on the basis of the generated reports. The results indicate that the services have common features such as log-on and log-off report, failure and log access reports that are established. It was also concluded that SOX compliance covers the reports of GLBA, HIPAA and PCI-DSS. However, variables such as privacy, security and user access were not mentioned, making the comparison table less accurate.

This paper [19] illustrates the seven most dangerous threats and how it affects cloud by relating the threats to the regulations. The assessment of compliance can be done on the basis of mapping and reference point. However, this paper doesn't highlight much on compliance and security. This research paper [20] reflects the study on cloud regulations. It states that in spite of all the advancements, inconsistencies still prevail in cloud environment. This leads to issues while assessing the security and privacy compliance in cloud. Moreover, it is mentioned that the basic requirements for certain regulations are privacy, integrity and security. Organizations will be legally responsible for any cases of data breaches. According to the author of [21], healthcare organizations that use COBIT are qualified for 50% of NIST standards. The author concluded that rising threats, shortage of privacy control experts, huge implementation and maintenance expenses are some challenges faced by the healthcare organizations. The author has advised that organizational compliance can be enhanced using the analysis of regulation overlaps and the best practices. Block diagrams indicate the overlaps; however, the kind and nature of overlap is not differentiable.

10.6   ACCESS-BASED CONTROL MECHANISMS IN HYBRID CLOUD ENVIRONMENTS

This section displays certain methods to control existing mechanisms in the hybrid cloud setting.

   i. An identity system is used to display the identity management model. For having access to the data, initially the user will have to identify using their credentials, and then a token will be issued by the identity system. This token is developed in such a way that it is refreshed on a timely basis to ensure that no unauthorized personal gets the data access. To have an access to the cloud services from an application, the application will have to undergo with a set of necessary variables. The cloud web server is able to identify the user through the token. This communication is carried out by the predefined metadata. After the successful verification of the token, cloud will provide access to the requested web services. The process is illustrated in Figure 10.2. This process has proven to be effective and safe in authorizing cloud web servers. It also has the advantages of Security Assertion Markup Language (SAML) and OAuth protocols.

  ii. A cloud web service comes with access management and attribute-based access control (ABAC). ABAC is used to monitor the users on the basis of various attributes. Some of them used here are environmental conditions, resources, location and policies. The main aim of this is to guarantee user authorization to have access to certain web services. ABAC-based authorization is enabled by a rule engine with priorly defined XML rules. These rules vary for each identity provider web service. The rule engine logs assist in providing system accountability. Independent rules define if a system is flexible and scalable. The algorithm functions by assessing the identity of

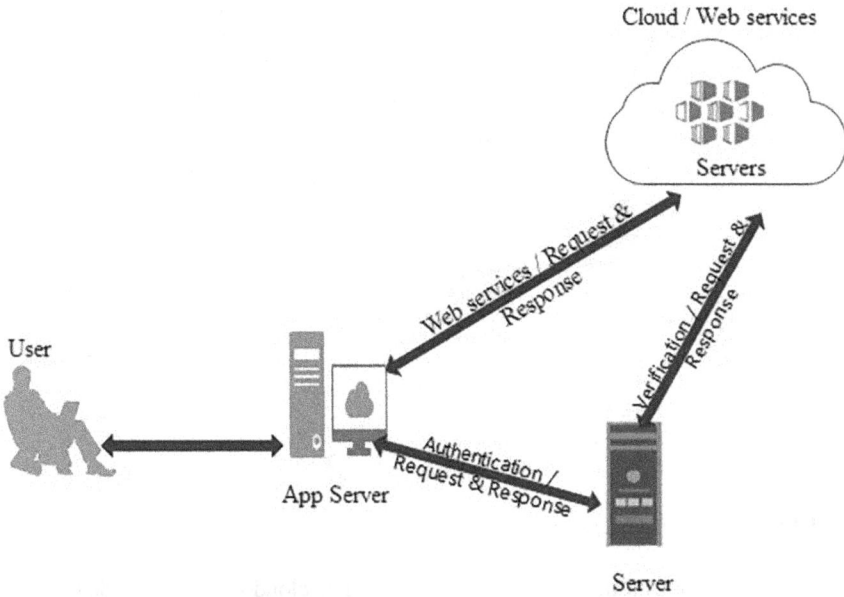

FIGURE 10.2   Managing identity.

the person and the resource time, organizational policies and the location. This is illustrated in Figure 10.3.

iii. In [23], the authors used a scheme known as cryptographic role-based access control, which enables the data to be available only for authorized users. It also highlights the problems caused by outsourcing, organizational policies, policy updates and problems raised by combining access controls and cryptography. A control model explained in [24] is known as role-based access control (RBAC). It is designed to have two mapping systems: user to role and role to privileges on data. This is an effective method of safely storing data in cloud. An encryption model known as role-based encryption (RBE) also ensures safety storage in cloud setting. The researchers in [25] described the application of RBAC to propose a scheme to organize and manage data exchange in cloud. To control the cloud policies, cryptographic RBAC combined with cryptographic controls was developed. Organizations that have highly sensitive data can apply trust models to store encrypted data.

10.7 MONITORING AND AUDITS IN A HYBRID CLOUD SYSTEM

Cloud system monitoring is conducted on the basis of tools that access servers, resources and applications. These tools mainly originate from the following.

1. **In-house tools from the cloud provider**: These tools are economic; these are incorporated within the service and do not need separate installation procedures.

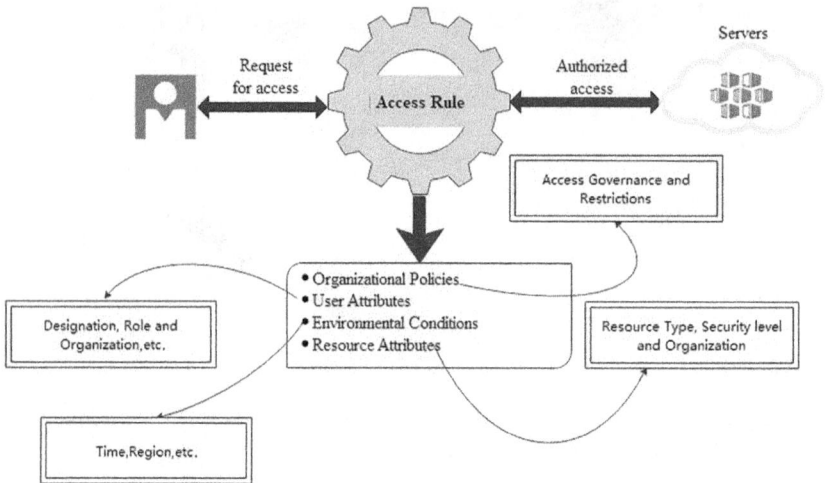

FIGURE 10.3   Managing access.

2. **Tools from independent SaaS provider:** The cloud services and SaaS may look different; however, the services function without any interruptions. The providers are also capable of performance management and expenses.

Cloud monitoring tools are responsible for evaluating the issues that hinder the services from reaching the customers. Data users are generally provided based on security, performance and the customer services.

- **Cybersecurity:** is an important procedure **to** safeguard networks from malicious attacks. IT professionals are competent to recognize the security breaches within the systems much earlier, which will help in preventing any damage to the data and mitigating the loss.
- **Regular monitoring practices:** Timely scheduled monitoring will benefit the companies by detecting the errors, and proper management of the errors will minimize the damage and the decline in the performance of function. This will lead to improving the user experience, thereby increasing the profits with customer retention.
- **Measurement of speed**: Parameters such as customer satisfaction can be measured using variables such as user experience and network function. However, the speed can be monitored and a database can be formed for the same that will eventually be used by the organization to update the related website and applications.

Timely scheduled monitoring in the companies will lead to obtaining a dataset with troubleshooting problems. Some of the benefits of cloud monitoring are as follows:

- Installation procedures will be uncomplicated as configurations and infrastructure are already built.

- Hardware is preserved safely and maintained by the host.
- The size of the solutions varies depending upon the organization; therefore, increased cloud activity will lead to seamlessly scaling the monitoring tool.
- The entire procedure of start-up will be cost-efficient as the subscription-based solutions will be affordable and maintenance cost will be divided among multiple users.
- Resources will not interfere with local issues as they are not a part of the server/workstation.
- **Application of multiple tools**: The ability to use multiple devices such as desktops, laptops and phones will help the company to carry out remote monitoring.

A cloud auditor is a third party that conducts independent monitoring of cloud services, thereby assessing the performance and providing unbiased opinions. The main aim of this audit is to evaluate the standards and evidences. The responsibility of a cloud auditor is to assess the security controls, privacy and performance provided by the cloud to its users. Audit stands out to be essential in accordance with the regulatory agencies. The IT system of a company is responsible for offering data safety measures such as technical, operational and managerial safeguards to protect the confidentiality and integrity of the data.

Security audit procedures involve assessing the security controls to see if it's used correctly. Also, the outcomes and its security requirements are examined. Furthermore, compliance with regulatory policies is also checked. For example, auditor must make sure that data be retained in accordance with the regulatory policies, leaving that data contents unaltered and safe.

Privacy audits are helpful in maintaining compliance between the regulatory agencies and the privacy laws. On the whole, the audit assures confidentiality and integrity of personal data at every developmental stage.

10.8 HARDWARE CLOUD SECURITY MODELS IN HYBRID ENVIRONMENTS

In view of the existing limitations, organizations that use cloud services are seeking for tools that can provide self-encrypting keys. In accordance with that, the hardware security model (HSM) is an effective solution. HSM is developed in a way that the encrypted keys are checked between the server and the digital world. Apart from creating and operating the encrypted keys, data such as firmware are organized by internal and external protocols. Usually, HSM is given a very limited access to organizational network. It operates on OS and has improved security and generates and stores randomized encryption keys. It is also capable of providing systems access in case unauthorized users attain access to certain data. Despite all the advantages, one limitation of HSM is that the physical structure makes it inappropriate for cloud businesses. Nevertheless, HSM still remains the most appropriate tool for companies that connect their IT structure to public domain and a part of hybrid cloud model and the expense toward this model could be high. For organizations that are relatively smaller and do not have their own IT infrastructure making it feasible for them to store data in cloud, HSM does not count as a tool of choice. Cloud services have made it possible for their udders to develop their own encryption keys and protocols using HSM, and even though HSM is managed by cloud, the users have control on the cryptographic keys. Once the keys are formed, they are securely placed in cloud environment, separate from public cloud. This step ensures that the keys remain unharmed even in case of unauthorized breach. With an attempt to offer added security, data centers known as collocation is maintained on-site to store the keys. HSM enables cloud users to access measurable security, allowing them to safeguard confidential data. Companies possessing strict compliance policies will be capable of managing their encrypted data.

In today's world, it has become an essential need to secure data on cloud. HSM services have greatly aided the cloud providers to provide customers with safe measures. Even though offering hundred percentage of data security is not ideally possible, developing and managing encryption keys by users and assessing protocols greatly add to making cloud a safe environment for storage.

10.9 CONCLUSIONS

As a lively measure to data security in hybrid cloud deployments, nowadays, there is an increasing trend for integrating the privacy and security standards with a cloud computing environment, for further processing and computation processes. This process facilitates operational and business activities. Yet, hybrid cloud computing infrastructures are mostly semi-trusted and it may subject to numerous security threats and vulnerabilities. Further, the hybrid clouds are generally resource-constrained in nature; as a result, it may lead to numerous performance- and efficiency-related overheads. Hence, this chapter aims to explore privacy and security compliance standards in hybrid cloud environments. In this context, various security threats and compliance models are discussed followed by access-based control mechanisms in hybrid cloud environments. Furthermore, monitoring and audits in a hybrid cloud system and hardware cloud security models in hybrid environments are deeply discussed.

REFERENCES

1. I. Indu and P.R. Anand, Identity and access management for cloud web services. *In 2015 IEEE Recent Advances in Intelligent Computational Systems (RAICS)*, IEEE, December 10 2015, pp. 406–410.
2. P.R. Anand and V. Bhaskar, A unified trust management strategy for content sharing in peer-to-peer networks. *Applied Mathematical Modelling*, 37(4): 1992–2007, 2013.
3. Z. Wang, D. Huang, Y. Zhu, B. Li, and C.J. Chung, Efficient attribute-based comparable data access control. *IEEE Transactions on Computers*, 64(12): 3430–3443, 2015.
4. H. Nicanfar, P. Jokar, K. Beznosov, and V.C. Leung, Efficient authentication and key management mechanisms for smart grid communications. *IEEE Systems Journal*, 8(2): 629–640, 2013.
5. J. Li, X. Huang, J. Li, X. Chen, and Y. Xiang, Securely outsourcing attribute-based encryption with checkability. *IEEE Transactions on Parallel and Distributed Systems*, 25(8): 2201–2210, 2013.
6. Z. Yan, M. Wang, Y. Li, and A.V. Vasilakos, Encrypted data management with deduplication in cloud computing. *IEEE Cloud Computing*, 3(2): 28–35, 2016.

7. Q. Liu, G. Wang, and J. Wu, Secure and privacy preserving keyword searching for cloud storage services. *Journal of Network and Computer Applications*, 35(3): 927–933, 2012.
8. R. Rajmohan, T. Ananth Kumar, M. Pavithra, and S. G. Sandhya. 11 Blockchain. In E. Golden Julie, J. Jesu Vedha Nayahi, Noor Zaman Jhanjhi (Eds) *Blockchain Technology: Fundamentals, Applications, and Case Studies*, p. 177. Boca Raton, FL: CRC Press, 2020.
9. Coresecurity IAM. [Online] https:www.coresecurity.comiam-products.
10. M. Kaufman, Data security in the world of cloud computing. *IEEE Security and Privacy*, 7(4), 61–64, 2009.
11. C. Cachinand and M. Schunter, A cloud you can trust. *IEEE Spectrum*, 48(12), 28–51, 2011.
12. T. A. Kumar, A. John, and C. Ramesh Kumar. 2 IoT technology and applications. *Internet of Things*, 43–62, 2020.
13. M. Ali, S. U. Khan, and A. V. Vasilakos, Security in cloud computing: Opportunities and challenges. *Information Sciences*, 305, 357–383, 2015.
14. H. Takabi, J. B. D. Joshi, and G. J. Ahn, Security and privacy challenges in cloud computing environments, *IEEE Security and Privacy*, 8(6), 24–31, 2010.
15. A. Juels and A. Opera, New approaches to security and availability for cloud data. *Communications of the ACM*, 56(2), pp. 64–73, 2013.
16. COBIT, IT Governance Framework - Information Assurance Control, ISACA. http://www.isaca.org/Knowledge-Center/cobit/Pages/Overview.aspx.
17. FedRAMP, Federal Risk and Authorization Management Program (FedRAMP). https://www.fedramp.gov/resources/documents/.
18. O. Mirković, Security - How to measure compliance. *MIPRO Proceedings*, 2008.
19. C.M.R. Silva, J.L.C. Silva, R.B. Rodrigues, L.M. Nascimento, and V.C. Garcia, Systematic mapping study on security threats in cloud computing. *IJCSIS*, 11: 3, 2013.
20. J. Ruiter and M. Warnier, *Computers, Privacy and Data Protection: An Element of Choice*, pp. 361–376. Berlin, Germany: Springer Science & Business Media, 2011.
21. B.M. Netschert, Information security readiness and compliance in the healthcare industry. Stevens Institute of Technology, 2008.
22. L. Zhou, V. Varadharajan, and M. Hitchens, Trust enhanced cryptographic role-based access control for secure cloud data storage. *IEEE Transactions on Information Forensics and Security*, 10(11): 1556–6013, 2015.
23. D. F. Ferraiolo and D. R. Kuhn, Role-based access controls, *In Proceedings of the 15th NIST-NCSC National Computer Security Conference*, NIST, National Computer Security Center, Baltimore, October 10–13, 1992, pp. 554–563.
24. L. Zhou, V. Varadharajan, and M. Hitchens, Integrating trust with cryptographic role-based access control for secure cloud data storage, in *TrustCom 2013*, IEEE, Melbourne, Victoria, Australia, July 2013, pp. 560–569.
25. L. Zhou, V. Varadharajan, and M. Hitchens, Achieving secure role- based access control on encrypted data in cloud storage. *IEEE Transactions on Information Forensics and Security*, 8(12), 1947–1960, 2013.

11 Threat Detection and Incident Response in Cloud Security

Satheeshkumar Rajendran
Rakuten Mobile

A. Valarmathi
University VOC College of Engineering

M. Satheesh Kumar
National Engineering College

CONTENTS

DOI: 10.1201/9781003219880-11

11.1 INTRODUCTION

Cloud computing is a network access technique that offers on-demand network access to a pooled collection of customizable computing resources, for example servers, networks, services, storage, and applications [1]. Cloud computing model can be instantly provided and delivered with minimum administration involvement or communication between service providers [2]. In order to provide cheap and easy access to externalized IT (information technology) services, cloud computing [3] as shown in Figure 11.1 has emerged as a common solution.

11.1.1 ATTRIBUTES OF CLOUD

It is considered that any cloud computing service will have these five characteristics [4] that are listed below.

    a. **On-demand self-service**: A user can autonomously supply computing capacity, such as database server and server time, as required without needing human intervention with the service provider.
    b. **Broad network access**: Features are offered in cloud computing through standard frameworks and through the network that allows diversified customer platforms such as cell phones, laptops, and PDAs.
    c. **Resource pooling**: To support multiple users, the computational resources of the provider are pooled using a multi-tenant model with different virtual and physical resources dynamically distributed and reassigned to customer demand accordingly.
    d. **Rapid elasticity**: Services may be supplied flexibly and simply, and, in certain cases, autonomously, to fast scale-out and fast scale-in.
    e. **Measured service**: Cloud systems optimize and evaluate resource utilization autonomously (e.g. processing, bandwidth, storage, and active user accounts) by employing a monitoring feature at an appropriate level of

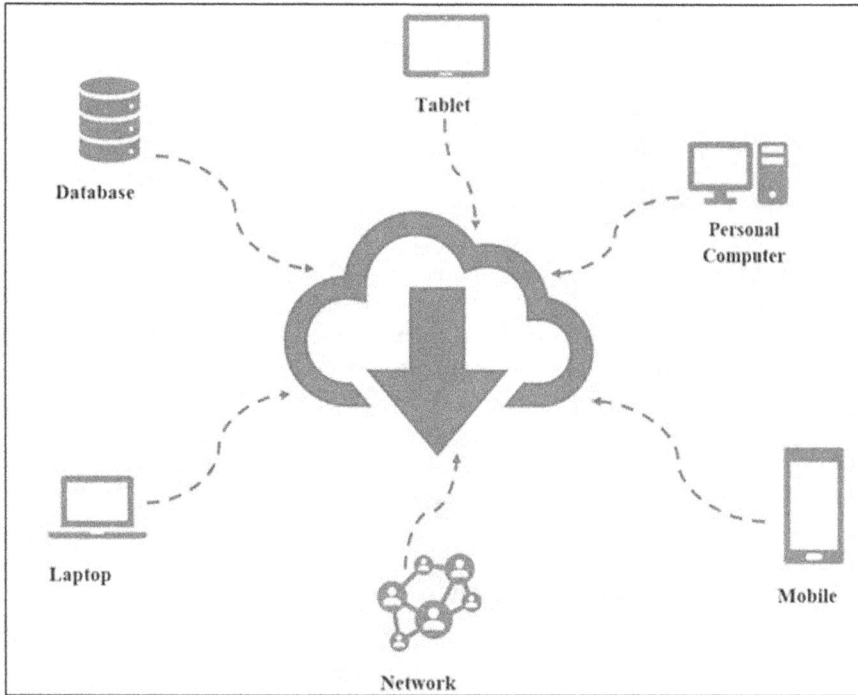

FIGURE 11.1 Cloud computing environment.

granularity for the different kinds of operation. The usage of resources may be tracked, managed, and disclosed, providing responsibility for both the provider and the customer of the service.

11.2 CLOUD DEPLOYMENT AND MODELS

Based on where the infrastructure of the environment is located, the deployment strategies are categorized as follows.

11.2.1 DEPLOYMENT STRATEGIES

a. **Public cloud**: The public cloud services [5] are defined as being accessible over the Internet to customers through a third-party service provider. While it may be reasonably economical or free to use, the word "public" does not necessarily imply free.

b. **Private cloud**: A private cloud [6] has many advantages, such as being scalable and service based, of a public cloud computing environment. The discrepancy between a public and a private cloud is that data and processes are handled within the enterprise in a private cloud-based service without the constraints of security exposure, network bandwidth, and legal needs.

   c. **Community cloud**: A cluster of companies with shared interests manages and uses a community cloud [7], which may include standard security criteria or a shared objective.

   d. **Hybrid cloud**: A hybrid cloud is a mixture of an interoperating public and private cloud [8]. Users usually outsource non-business-critical information and processing to the public cloud in this model, while retaining control of business-critical resources and data.

The goal of this chapter is to provide a broad and complete overview of threat detection in cloud security and to offer a comprehensive analysis of incident response in cloud security. The remainder of this chapter is structured as follows: Section 11.2 contains cloud deployment and models. In Section 11.3, we provide a detailed illustration of cloud computing and security. In Section 11.4, we describe threat detection in cloud. In Section 11.5, we explain the incident response in cloud. Before summarizing our work in Section 11.7, we discuss top cloud security providers in Section 11.6. Finally, the future scope is narrated in Section 11.8.

11.2.2 MODELS FOR CLOUD DELIVERY

It is required to know what kind of models can be deployed in the organization based on the different levels of control, flexibility, and management. The following classification will give a clear-cut picture on the cloud computing models.

   a. **Software as a service (SaaS)**: In an era of cloud computing, SaaS is software [9] that one or more providers own, distribute, and operate remotely and that is delivered in a pay-per-use manner.

   b. **Platform as a service (PaaS)**: As a service, this form of cloud computing [10] presents a development environment. The customer may use the equipment of the middleman to create their own software and distribute it through the Internet and servers to the customers.

   c. **Infrastructure as a service (IaaS)**: Infrastructure as a service provides an outsourced service [11] for platform virtualization. As a service, the user will monitor the environment. Clients can obtain resources such as a complete operating system, memory, application programs, and probable infrastructure elements such as firewalls rather than software, servers, access points, or data centres.

11.3 CLOUD COMPUTING AND SECURITY

In this digital world, with the increasing innovations over Internet, the demand and need for the usage of servers, databases, and networks have increased for reasons such as data storage and processing of the data. With the increasing volume of data flowing [12] through the Internet across these servers and networks, a need has raised for virtualization. And with virtualization, companies [13] have started processing and storing sensitive data of their own and their customers in the cloud environments.

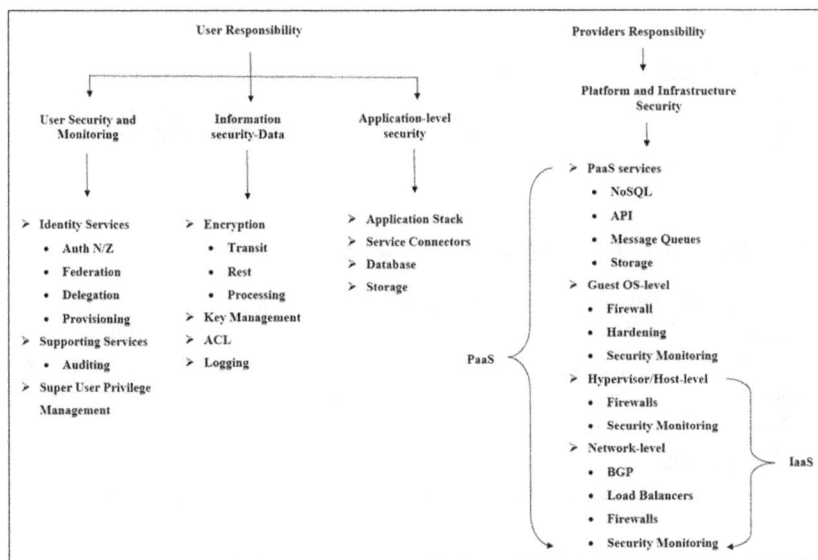

FIGURE 11.2 Cloud security architecture. (Source: InfoQ.)

At the corresponding moment, these sensitive data can be exploited [14] if they are not properly secured in the cloud environment.

Cloud service providers [15] have various activities in setting up these cloud environments for their cloud customers, and nowadays, as a regular part of their activity, most of the cloud providers are offering "security" to their customers as an added service. As depicted in Figure 11.2, in cloud mostly security [16] is a collective responsibility of both the cloud service provider and the cloud consumer that is the organization. Mostly, "privacy" will always be included in this security service. Most of the common security features listed [17] under these services by the cloud service providers and the organization are the following:

1. Authentication/authorization
2. Identity and access management
3. Data integrity/availability/confidentiality
4. Cyber threat intelligence
5. Monitoring (apps/users/assets)
6. Incident response and management
7. Managing policies (firewall/WAF/rules)
8. Data privacy (e.g. GDPR).

11.3.1 AUTHENTICATION/AUTHORIZATION

Authentication [18] is a unique way of allowing a user to have access over the information stored in the system, whereas authorization prevents the unauthorized user access over the information stored in the system.

11.3.2   IDENTITY AND ACCESS MANAGEMENT

Mostly, organizations which have a large number of enterprise customers will be using this identity and access management [19]. These are a set of policies/rules defined mostly in a tool to measure/ensure the correct level of authorized access to the company's data and resources (e.g. Cisco IAM).

11.3.3   DATA INTEGRITY/AVAILABILITY/CONFIDENTIALITY

This refers to the level of access, security, and privacy a user has in the company's data and resources [20]. For example, a normal user will have access only to the information that they are allowed to view/access; unauthorized access or elevated access is prohibited, and this is ensured by availability. A power user will have access for making authorized changes in the data within the system; other users and public access/changes to the data will be prohibited, and this is ensured by data integrity. Prohibition of unauthorized access to the data and making changes to the data is ensured by confidentiality.

11.3.4   CYBER THREAT INTELLIGENCE

This is a generalized service that we can use based on an organization's need. A cyber threat intelligence professional will collect data across the Internet about attacks/indicators of compromise/attacker/attack vector and mechanism specific to a cloud environment and will share it to the monitoring and incident response team [21]. The report will also be having the necessary action items to be taken within the cloud environment to prevent the attacks. This is mostly an advisory service to the security monitoring team which uses the SIEM tools and the incident response team.

11.3.5   MONITORING

In cloud environment, monitoring is an important aspect in terms of providing security by means of a service. In a cloud environment [22], multiple virtual hosts will be running and a large number of user bases will have access to these connected hosts and the apps running within the hosts. So, from the security point of view, we should have a track on all the hosts, apps running, and even the users accessing these apps and hosts and their activities. Organization uses SIEM tools for security monitoring; we are going to see it in depth later in Section 11.4.

11.3.6   INCIDENT RESPONSE

Incident response is very crucial not just for cloud environment, but for any organization who has sensitive data to protect. Normally, an incident is a reference [23] to a compromised system. Once a system is compromised, it's the role of incident response and management teams to create a bridge to connect various teams to seal the data leakage and find the indicators of compromise and eradicate them within the system. During such incidents, the threat intelligence reports will be useful in narrowing down the indicators of compromise.

11.3.7 Managing Policies

This refers to the configuration part, where a set of rules and policies [24] are created for the organization to access the cloud environment. This is basically a defined guideline which enables the company to have/maintain control over its owned resources and its associated accesses.

11.3.8 Data Privacy

This ensures that the customer's shared data [25] will be used only for the purpose they are intended to be. This differs from geographical reasons as this needs to comply with the local government's defined rules of how the customer's data can be collected and used.

11.4 THREAT DETECTION IN CLOUD

In cloud, threats are detected with the help of security team monitoring the entire cloud including the number of hosts linked and the number of web apps/services mounted in the virtual machines and the user bases who have access to the apps/ services and the hosts and also the network traffic flow [26]. There are plenty of tools available for monitoring the cloud infrastructure and environment. It depends on the organization to select one and use it based on its needs. Splunk tool is one such example, as shown in Figure 11.4. Tech giants such as Microsoft and Google have their own internal tools for monitoring their cloud apps and environment for threat detection, as shown in Figure 11.3. Threats are detected in the cloud by the prewritten threat identifier rules and threat scenario rules developed by the security team and architects.

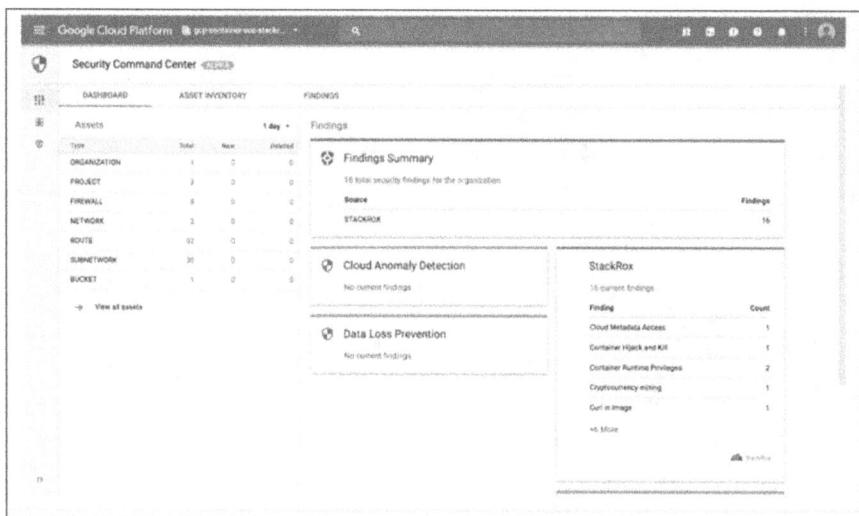

FIGURE 11.3   Google Cloud Platform. (Source: Google Cloud Platform.)

11.4.1 Event-Based Threat Detection

The alerts triggered under the name "threat matches" in Figure 11.3 are based on a specific set of threat identifier rules derived by the security team [27]. These rules are written and added into the SIEM (security information and event management) tool. SIEM tool will have the logs of all the events occurred in the cloud, as the entire cloud infrastructure and the apps/users/services within are integrated with the SIEM tool. It is the responsibility of the security assurance team in an organization. Some examples of the event-based threat detections are as follows:

1. Brute-force attack
2. Abusing identity and access management.

11.4.1.1 Brute-Force Attack

When an adversary tries for an unauthorized access and gains success after repeated failed log-ins into the system, it is termed as a brute-force attack [28]. For example, a legitimate user forgets his password and he uses repeated wrong passwords (let's say eight failed attempts) and then succeeds on the ninth attempt after entering the correct password. It will be considered as a brute-force attack by the SIEM tool rule, as the rule defines a maximum number of failed log-ins as the threshold and triggers the alert when the number of attempts pass the limit.

As a result of the alert, the security monitoring team have to triage the alert and get confirmed with the user whether it is a legitimate try or if the log-in attempts are made without the user's knowledge. If the user denies the failed log-in attempts, then it should be considered as a brute-force attack and the user's account needs to be disabled immediately and the security team should start the investigation going through the linked logs and events (Figure 11.4).

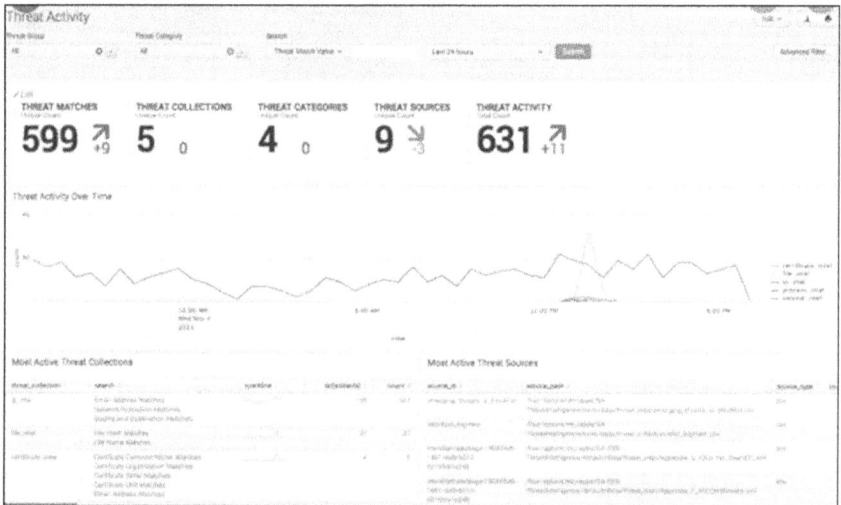

FIGURE 11.4 Splunk threat detection alerts. (Diagram reference: Splunk.)

11.4.1.2   Abusing Identity and Access Management

IAM tools are strictly configured for the admin users to add/remove a user account within the company's internal network and not from any outside network connection [29]. When such a user account added outside of the company's network was detected through the event log monitoring alert system, it will be considered as an IAM abuse case and appropriate actions will be taken immediately to prevent further compromise to the internal network.

11.4.2   Use Cases

There are multiple other use cases that can be written/developed and configured in the SIEM tools for threat detection specific to possible data loss and data leakage in cloud. Some of the examples are listed below for reference.

a. **Authentication**: By authenticating the allowed limit or user to access data [30], we can prevent data loss or leakage to external sources.
   i. **Multiple failed log-in attempts from the same source**: This can block or trigger an alert for the security professionals when the number of failed log-in attempts from a source exceeds the allowed number of failed log-in attempts.
   ii. Source account inactive for more than the allowed number of days.
b. **Sneak attack through VPN**: VPN connections are the most common way of misuse, and hence security professionals need to monitor the remote user's connectivity locations and their access over data.
c. **Advantage or abuse of privileged user access**: Root access or the admin access is given to certain users and such privileged access can be misused for the access of data and, hence those privileged access need to be monitored.
d. **Authorization**:
   i. **Unauthorized data** access: This can be monitored by looking for systems/users who are having suspicious activity or suspicious applications running in the system.
   ii. **Usage of disabled accounts**: Disabled accounts, for example inactive accounts or accounts of users who are no longer a part of the organization, need to be checked for any activity for a specific period of time.
   iii. **Geographic locations**: Accounts or any connection requests coming from a suspicious location such as out of business coverage location or out of consumer location need to be monitored for suspicious unauthorized access.
e. **Malware lookout**: Analysing the trend of the incoming malware or suspicious files [31] both within the organization and outside the organization using the threat intelligence trend report.
f. **Detecting the attackers**: In case of an attack, the first thing to be noticed or monitored is the location from where the attack is initiated and the attack's focus on data. This will help in detecting the actual attackers behind the attack.

g. **Vulnerability scanning and management**: In cloud, the first and foremost concentration for detecting compromises to have an up-to-date knowledge of the available vulnerabilities of the cloud environment [32] and the associated assets including the ports, network devices, services, and the technologies.

11.5 INCIDENT RESPONSE IN CLOUD

An incident is referred to as a compromise in the system in security aspects, which usually results in data theft, data loss, and revenue loss if the incident affects the production. And incident response is referred to as an approach in which the compromised organization will prepare, detect, contain, and recover from the security breach. Figure 11.5 represents the normal incident response cycle as per the NIST standards [33].

11.5.1 IMPORTANCE OF INCIDENT RESPONSE

Incident response limits the amount of data loss or data theft in a security breach within the organization [34]. Incidents need to be mitigated as quickly as possible to limit the damage to the organization. It is necessary to understand the primary goals of incident response in terms of mitigating the attacks.

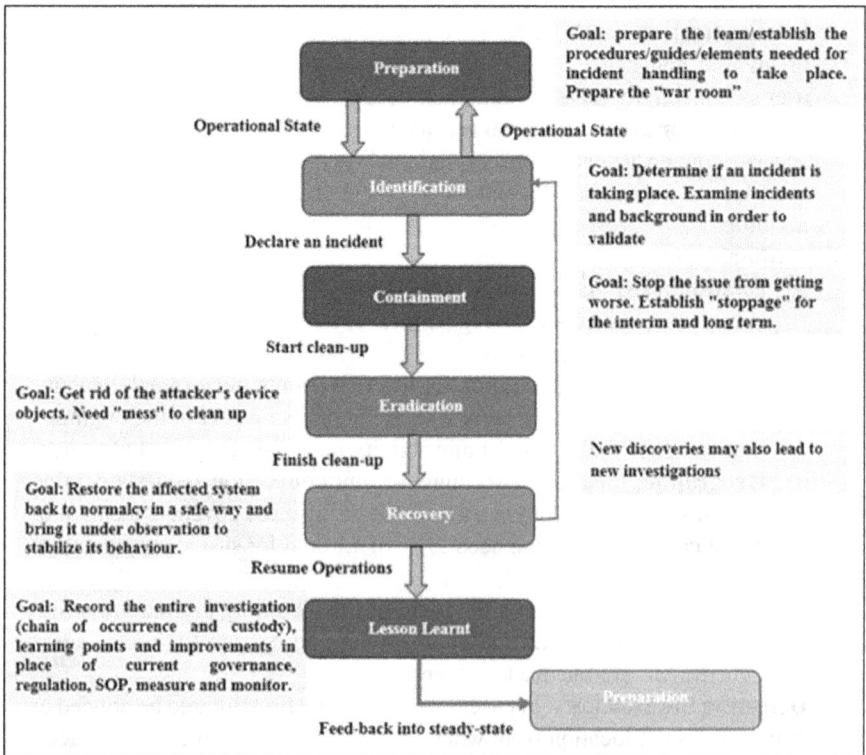

FIGURE 11.5 Incident response goals. (Diagram reference: Experts Exchange.)

The primary goals of incident response are given in Figure 11.5. The purpose of the planning stage is to ensure that the company can respond to an event comprehensively at a moment's notice. The IR team would need to identify threats [35] from log warnings, IDS/IPS, firewalls, and any other unusual behaviours on the network during the identification stage.

Once a threat has been detected, according to the policy developed during the planning process, it should be reported and communicated. The IR team should work to mitigate the danger after a threat has been detected to avoid more harm to other systems and the company at large. Extermination is a stage of effective incident response that entails removing the danger and restoring impacted systems to their previous condition, ideally with little data loss.

In order to allow normal operations to resume, the recovery process of an incident response plan [36] includes restoring all damaged systems and equipment. Before systems are backed up and running, however, it is important to ensure that the cause of the breach has been identified to prevent another breach from occurring again. Following the resolution of a safety event, a lesson learned session takes place. This involves analysing how successfully the incident response plan worked to fix the problem and indicating any modifications that need to be made.

11.5.2 INCIDENT RESPONSE PLANNING

As shown in Figure 11.6, whenever an event occurs, the identification of the nature of the event is carried out. The incident response plan [37] in accordance with the event is launched, and the required actions are performed. Based on the result, the next steps are decided for disaster recovery.

11.5.3 INCIDENT HANDLING IN CLOUD

An incident is an inadvertent disruption in an IT operation or a decrease in quality. Traditional IT service management (ITSM) is dependent on different software

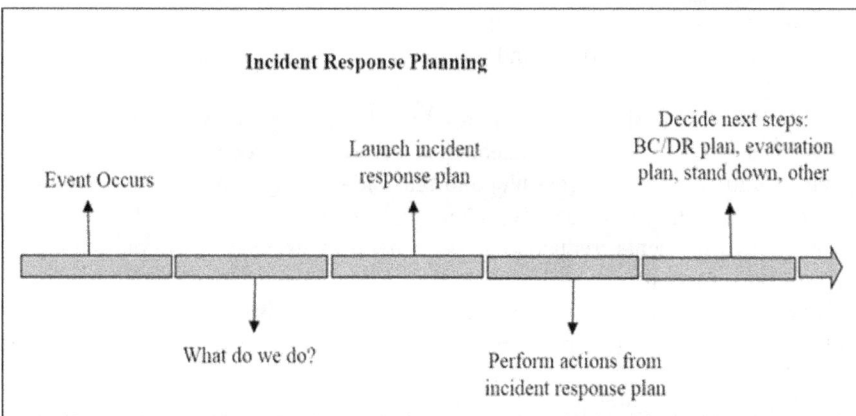

FIGURE 11.6   Incident response planning. (Diagram reference: searchdisasterrecovery.com.)

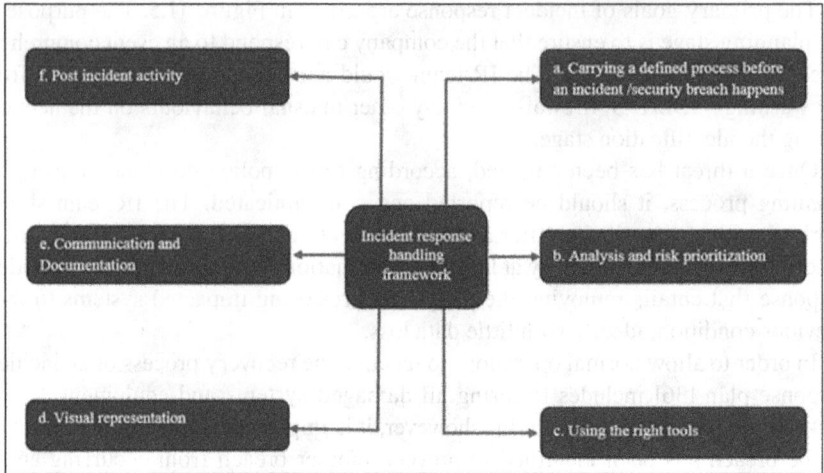

FIGURE 11.7 Incident response handling framework.

and systems to track, examine, and warn the teams to events as they develop [38]. Incident response in cloud is no different than the other environments; the impact of any security breach can be minimized as long as the following steps depicted in Figure 11.7 are strictly followed.

11.5.3.1 Carrying a Defined Process before an Incident/Security Breach Happens

There is no way to foresee any kind of event or circumstance to fix. It is necessary to be prepared, however.

- Create playbooks of common incident management procedures.
- Handle incidents rapidly.
- Enhance the internal and external communication.
- Reduce revenue losses as much as possible.
- Promote learning process and progress.

Define the scripts that can be followed by the team to communicate about crucial outages with clients and stakeholders. Verify the processes are periodically updated (automated where possible) and that the training manuals are available to team members. Having a recovery plan in place means it can rapidly and confidently resolve incidents, reducing the likelihood of expensive miscommunication and misunderstanding.

This process needs to have a clear picture of what needs to be done when an incident happens. It can be referred to as an incident management plan, and it helps in

- Resolving the incidents very quickly
- Setting up a bridge call improving the communication between the internal environment and the dependent external environment

- Minimizing the data and revenue loss
- Focusing on the learning of the trends of the security breach which results in improvement.

11.5.3.2   Analysis and Risk Prioritization

When an incident happens, it is necessary to make fast choices. What's the issue? What hazards does this pose? What hazards are most important to tackle? Who is it affecting? These questions need to be addressed when it is necessary to minimize the risks and reduce business interruptions [39]. Using key monitoring systems, escalation, and diagnostic processes to evaluate the response, assess the effects, and prioritize risks. Consistent communication channels between team members, as well as outlined accountability criteria, are ensured.

The below-listed ones are few of the existing security risks in a cloud environment. Data loss/data leakage is the most common security risk in cloud computing.

- Data breaches and malware infection
- Denial-of-service attack
- Hijacking accounts
- Insider threats
- Insecure applications
- Identity thefts
- Potential revenue loss.

After an incident, an IR member should make decisions very quicker analysing the below questions.

- The problem and the risk posed by the incident.
- The impact on the risks and the associated users or customer base.

Making quicker decision for these questions under a pressurized situation will result in quicker mitigation of the incident, minimizing business disruption. Also, a proper communication channel or call needs to be established involving all parties of the business, developers, and stakeholders to identify the indicators of compromise and mitigate them early. And in case of doubts in taking these quicker decisions, without a second thought, it needs to be escalated to higher levels with proper findings and reports.

11.5.3.3   Using the Right Tools

The architecture of the cloud is also wide and complex, with several moving parts to control and track. Therefore, investing in the correct incident management tools to help the cloud incident response processes is critical.

Using the right tools will always help in narrowing down the incident [40], for example SIEM tool such as Splunk. Splunk is commonly used across various organizations for monitoring the security operations. For a cloud environment, logs of all the assets and users are integrated into Splunk and these logs are categorized based on the events and the alert dashboards. During an incident, Splunk helps in identifying the number of events and the users associated in them.

11.5.3.4   Visual Representation

The cloud infrastructure and environment are very complex, resulting in security incidents going undetected. To prevent such incidents going undetected, proper visualization [41] needs to be done defining the entire process in diagrams and mapped to every team and user associated in the cloud environment. There are many third-party tools available in the market for the visual representation of the cloud infrastructure.

11.5.3.5   Communication and Documentation

During an incident response activity, it is very important to have the proper communication with accurate documentations, especially the timestamps. Timestamps of every activity in an incident response plan help in deriving the proper root cause analysis and incident mitigation report. Also, many tools available for tracking and creating the incident reports. Few things that are most needed in an incident report are the following:

- Incident timestamp
- Person who handles the incident
- Activity history and timestamp
- Actions on every stage of analysis.

11.5.3.6   Post-Incident Activity

Once an incident is mitigated, we need to have a proper track of the incident in a centralized repository. This will help in narrowing down the future incidents. The reports will help in identifying the indicators of compromise and the appropriate actions taken for mitigating the same in case of similar incidents in future.

11.6   TOP CLOUD SECURITY PROVIDERS

Cloud security covers a wide variety of expertise, guidelines, software, regulations, and software packages used to safeguard cloud computing's virtualized IP, applications, services, records, and related infrastructure. It refers to network security, computer security, and more generally sub-domain of information security. The most common cloud providers are listed below.

11.6.1   Amazon Web Service (AWS)

It was the first to provide the infrastructure of cloud computing as a service in 2008 [42]. To understand the basics of responding to security incidents inside cloud setting, AWS has recently published the AWS Security Incident Response Whitepaper.

The whitepaper discusses how to plan the company for the identification and response of security incidents, explores the controls and resources available, offers up-to-date examples, and outlines remediation strategies that incorporate automation to enhance the speed of response.

AWS users inside an enterprise must have a clear interpretation of the procedures of security incident response [43], and security workers need to thoroughly understand how to respond to security problems.

Although education and training are key components of this, in order to iterate and refine processes, clients are allowed to exercise these abilities through simulations.

The foundation of an effective cloud incident response program is listed below.

a) **Educate**: Educate the security and incident management teams about cloud technology and how they are meant to be used by the company.

b) **Prepare**: Allow investigation skills and ensure full exposure to the tools and cloud resources accessible to the incident response team so they can locate and react to cloud incidents.

c) **Simulate**: Simulate both planned and unpredicted security incidents within the cloud environment to understand the efficacy of planning.

d) **Iterate**: Iterate to increase the size of response posture on the outcome of simulation, minimize delays, and further reduce risk.

11.6.2　MICROSOFT AZURE

The incident response process in Microsoft Azure [44] has the following steps:

a) **IR-1**: Planning – revise incident response procedure for Azure.
   i. Verify that the organization has procedures in place to respond to security issues, that these processes have been updated for Azure, and therefore that these processes are regularly exercised to guarantee preparedness.

b) **IR-2**: Planning – format the incident warning.
   i. In Azure Security Center, contact information has to be configured for security incidents.
   ii. If the Microsoft Security Response Center (MSRC) identifies a vulnerability, Microsoft utilizes this information to contact the appropriate client if an illegal or unauthorized party has accessed the client's data. The client also has options to configure the warning and notification of incidents in various Azure services based on the requirements for incident response.

c) **IR-3**: Discovery and evaluation – based on high-quality alarms, create incidents.
   i. Ensure about the mechanism for producing notifications of high quality and assessing warning quality.
   ii. This helps to learn lessons from previous events and send analysts priority warnings to avoid wasting time on false positives.

d) **IR-4**: Discovery and evaluation – examine an incident.
   i. Confirm that researchers may query and use multiple information resources to create a complete picture of what occurred when investigating possible incidents.
   ii. In order to prevent blind spots, different records ought to be gathered to monitor the actions of a possible intruder through the kill chain.

e) **IR-5**: Discovery and evaluation – select incidents based on priority.
   i.   Give analysts insight as to which events should be prioritized based on the severity of the alert and the susceptibility of the assets.
   ii. Each warning is given a severity by the Azure Security Center to order which warnings should be inspected first.
   iii. The severity is determined by the security centre's conviction in the warning finding or the logic used to generate the alert, as well as the amount of trust that the behaviour that triggered the alert was malicious.
f) **IR-6**: **Containment**, eradication, and recovery – automate the incident handling.
   i.   Automate the manual routine operations to improve the reaction time and relieve the load on analysts. Manual activities take longer to accomplish, which slows down each event and limits the number of events an analyst can handle.
   ii. Analyst tiredness is increased by manual work, which increases the chance of human mistake creating delays, which reduces analysts' ability to focus successfully on complicated jobs.
   iii. Use the workflow automation tools in Azure Security Center and Azure Sentinel to systematize the behaviour, or execute a script to respond to the received security warnings.

11.6.3 GOOGLE CLOUD

Each data event is unique, and the purpose of the data incident response process [45] is to protect the information of consumers, restore normal operation as quickly as possible, and satisfy both regulatory and contractual enforcement requirements [45].

The process of incident response software from Google is shown in Figure 11.8.

11.7 SUMMARY: CHALLENGES IN CLOUD SECURITY INCIDENT RESPONSE

The public cloud solves many of the problems that have plagued security activities in the past, but it also poses new issues that security professionals must learn to deal with.

Threat actors view cloud applications as high-value targets that contain sensitive information such as employee personal identity information (PII), data from vendors and customers, and financial and company data.

In the current real-world scenario, among SaaS, PaaS, and IaaS models, understanding the shared responsibility model of SaaS applications is important. In SaaS, both the client and the cloud service provider are responsible for the security against threat. End protection, data ownership, access management, application controls, network controls, host infrastructure, and physical security are the various responsibilities of the incident response system in SaaS cloud architecture. The client is accountable for the protection of the files, end station, account, entry, and individuality in the SaaS distribution model, while the cloud service provider is liable for the other components.

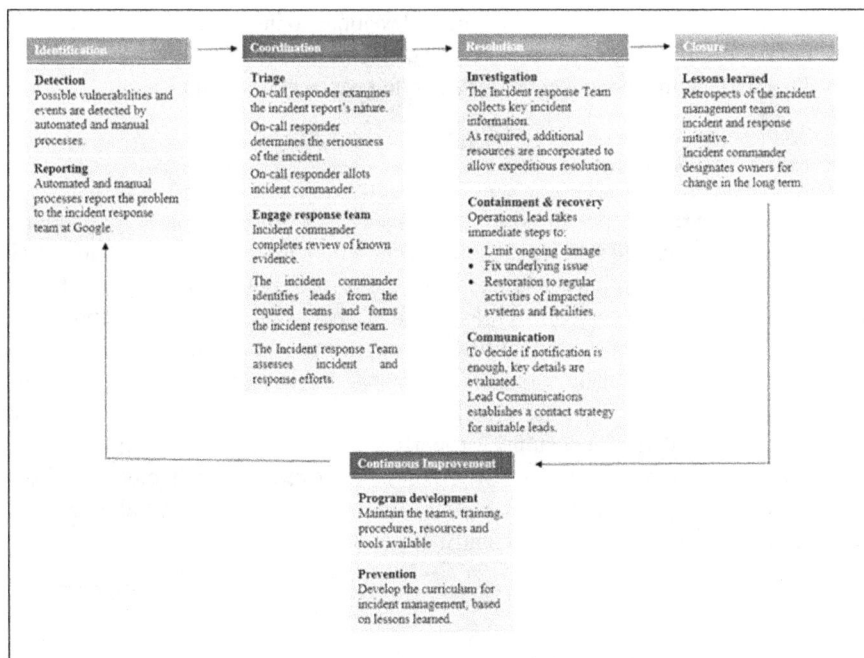

FIGURE 11.8   Incident response process in Google Cloud.

From the standpoint of incident response, organizations using a cloud framework can create a plan to track, manage, and react to cybersecurity events that affect the cloud architecture components under their accountability. These organizations can also implement a framework for evaluating cloud computing services and ensuring that companies possess the necessary skills to deal with cyberspace security issues in their area of influence.

Hence, it is required to manage the cloud security incidents in the context of the company's responsibilities and cloud service provider's responsibilities.

In scope of the organization's responsibilities, it is essential to contemplate the following factors for managing the cloud cybersecurity incidents.

1. For detecting suspicious activities that involves data, identities, or access, the organizations may use audit and access logs functionality to achieve the visibility of inadvertent accesses, data exfiltration, and data intrusion operations. In such scenarios, cloud application that provides API can be included in the framework of the logging technique, since intruders can utilize them to carry out harmful acts.

2. Company should consider if the cloud application logs provided can be integrated into their SIEM systems for detecting suspicious actions. Monitoring practices can then be incorporated within such tools by creating use cases and warnings for typical security issues involving records, identities, or access.

3. For incident response, protocols can be developed to inspect and resolve security issues involving application design elements under the company's control. This will aid in having a coherent and effective response to incidents.

11.8 FUTURE SCOPE

The generic incident response addresses issues such as out-of-date applications, proper firewall setup, and successful authorization policies and procedures and verifying that all platform activity is tracked and authorized. But the public cloud brings several roles to security teams, such as learning new skills, monitoring ephemeral assets, handling decentralized identities, and securing data maintained using cloud provider infrastructure services instead of servers. Because of the blind holes generated when utilizing the public cloud, adversaries can exploit and compromise essential and confidential consumer properties. This kind of compromises could jeopardize a company's activities and credibility. To protect their cloud environments, information techs must learn new technologies and expertise, but many lack the time and resources to do so efficiently.

REFERENCES

1. L. Wang, G. Von Laszewski, A. Younge, X. He, M. Kunze, J. Tao, and C. Fu, "Cloud computing: a perspective study," *New Generation Computing*, vol. 28, no. 2, pp. 137–146, 2010.
2. A. O. Akande, N. A. April, and J. P. Van Belle, "Management issues with cloud computing," in *ACM International Conference Proceeding Series*, pp. 119–124, 2013. doi: 10.1145/2556871.2556899.
3. K. Suresh Kumar, A. S. Radha Mani, S. Sundaresan, and T. Ananth Kumar, "Modeling of VANET for future generation transportation system through edge/fog/cloud computing powered by 6G." In: G. Singh, V. Jain, J. M. Chatterjee, L. Gaur (Eds) Cloud and IoT-Based Vehicular Ad Hoc Networks, pp. 105–124. Hoboken, NJ: John Wiley & Sons, 2021.
4. B. Furht, "Cloud computing fundamentals," In: B. Furht and A. Escalante (Eds) *Handbook of Cloud Computing*, pp. 3–19. Berlin, Germany: Springer US, 2010.
5. P. Hofmann and D. Woods, "Cloud computing: The limits of public clouds for business applications," *IEEE Internet Computing*, vol. 14, no. 6, pp. 90–93, 2010, doi: 10.1109/MIC.2010.136.
6. L. Qian, Z. Luo, Y. Du, and L. Guo, "Cloud computing: An overview," in Lecture Notes in Computer Science (including subseries Lecture Notes in Artificial Intelligence and Lecture Notes in Bioinformatics), vol. 5931 LNCS, pp. 626–631, 2009, doi: 10.1007/978-3-642-10665-1_63.
7. A. Marinos and G. Briscoe, "Community cloud computing," in Lecture Notes in Computer Science (including subseries Lecture Notes in Artificial Intelligence and Lecture Notes in Bioinformatics), vol. 5931 LNCS, pp. 472–484, 2009, doi: 10.1007/978-3-642-10665-1_43.
8. M. S. Aktas, "Hybrid cloud computing monitoring software architecture," *Concurrency and Computation: Practice and Experience*, vol. 30, no. 21, p. e4694, 2018, doi: 10.1002/cpe.4694.
9. W. T. Tsai, X. Y. Bai, and Y. Huang, "Software-as-a-service (SaaS): Perspectives and challenges," *Science China Information Sciences*, vol. 57, no. 5, pp. 1–15, 2014, doi: 10.1007/s11432-013-5050-z.

10. D. Beimborn, T. Miletzki, and S. Wenzel, "Platform as a service (PaaS)," *Business & Information Systems Engineering*, vol. 3, no. 6, pp. 381–384, 2011, doi: 10.1007/s12599-011-0183-3.
11. S. S. Manvi and G. Krishna Shyam, "Resource management for Infrastructure as a Service (IaaS) in cloud computing: A survey," *Journal of Network and Computer Applications*, vol. 41, no. 1. Academic Press, pp. 424–440, 2014, doi: 10.1016/j.jnca.2013.10.004.
12. I. A. T. Hashem, I. Yaqoob, N. B. Anuar, S. Mokhtar, A. Gani, and S. Ullah Khan, "The rise of 'big data' on cloud computing: Review and open research issues," *Information Systems*, vol. 47. Elsevier Ltd, pp. 98–115, 2015, doi: 10.1016/j.is.2014.07.006.
13. N. Jain and S. Choudhary, "Overview of virtualization in cloud computing," September 2016, doi: 10.1109/CDAN.2016.7570950.
14. R. Rai, G. Sahoo, and S. Mehfuz, "Securing software as a service model of cloud computing: Issues and solutions," *International Journal on Cloud Computing: Services and Architecture (IJCCSA)*, vol. 3, no. 4, pp. 1–11, 2013, doi: 10.5121/ijccsa.2013.3401.
15. R. Velumadhava Rao and K. Selvamani, "Data security challenges and its solutions in cloud computing," *Procedia Computer Science*, vol. 48, no. C, pp. 204–209, 2015. doi: 10.1016/j.procs.2015.04.171.
16. M. Almorsy, J. Grundy, and I. Müller, "An analysis of the cloud computing security problem," September 2016, Accessed: March 17, 2021. Online. Available: http://arxiv.org/abs/1609.01107.
17. K. Hashizume, D. G. Rosado, E. Fernández-Medina, and E. B. Fernandez, "An analysis of security issues for cloud computing," *Journal of Internet Services and Applications*, vol. 4, no. 1, pp. 1–13, 2013, doi: 10.1186/1869-0238-4-5.
18. N. M. Gonzalez, M. A. T. Rojas, M. V. M. da Silva, F. Redígolo, T. C. M. de Brito Carvalho, C. C. Miers, M. Näslund, and A. S. Ahmed, A framework for authentication and authorization credentials in cloud computing. *In 2013 12th IEEE International Conference on Trust, Security and Privacy in Computing and Communications*, 2013, pp. 509–516, IEEE.
19. I. Indu, P. M. R. Anand, and V. Bhaskar, "Identity and access management in cloud environment: Mechanisms and challenges," *Engineering Science and Technology, an International Journal*, vol. 21, no. 4. Elsevier B.V., pp. 574–588, 2018, doi: 10.1016/j.jestch.2018.05.010.
20. A. Tchernykh, U. Schwiegelsohn, E. ghazali Talbi, and M. Babenko, "Towards understanding uncertainty in cloud computing with risks of confidentiality, integrity, and availability," *Journal of Computational Science*, vol. 36, p. 100581, 2019, doi: 10.1016/j.jocs.2016.11.011.
21. N. Moustafa, E. Adi, B. Turnbull, and J. Hu, "A new threat intelligence scheme for safeguarding industry 4.0 systems," *IEEE Access*, vol. 6, pp. 32910–32924, 2018, doi: 10.1109/ACCESS.2018.2844794.
22. A. Patrascu and V. V. Patriciu, "Beyond digital forensics. A cloud computing perspective over incident response and reporting," in *SACI 2013–8th IEEE International Symposium on Applied Computational Intelligence and Informatics, Proceedings*, 2013, pp. 455–460, doi: 10.1109/SACI.2013.6609018.
23. D. Zissis and D. Lekkas, "Addressing cloud computing security issues," *Future Generation Computer Systems*, vol. 28, no. 3, pp. 583–592, 2012, doi: 10.1016/j.future.2010.12.006.
24. C. M. DaSilva, P. Trkman, K. Desouza, and J. Lindič, "Disruptive technologies: A business model perspective on cloud computing," *Technology Analysis & Strategic Management*, vol. 25, no. 10, pp. 1161–1173, 2013, doi: 10.1080/09537325.2013.843661.
25. A. Dutta, G. U. O. Chao Alex Peng, and A. Choudhary, "Risks in enterprise cloud computing: The perspective of it experts," *Journal of Computer Information Systems*, vol. 53, no. 4, pp. 39–48, 2013, doi: 10.1080/08874417.2013.11645649.

26. S. Krishnaveni and S. Prabakaran, "Ensemble approach for network threat detection and classification on cloud computing," *Concurrency and Computation: Practice and Experience*, vol. 33, no. 3, p. e5272, 2021, doi: 10.1002/cpe.5272.

27. A. Ambre and N. Shekokar, "Insider threat detection using log analysis and event correlation," *Procedia Computer Science*, vol. 45, no. C, pp. 436–445, 2015. doi: 10.1016/j.procs.2015.03.175.

28. A. Patil, A. Laturkar, S. V. Athawale, R. Takale, and P. Tathawade, "A multi-level system to mitigate DDOS, brute force and SQL injection attack for cloud security," in *IEEE International Conference on Information, Communication, Instrumentation and Control, ICICIC* 2017, February 2018, vol. 2018, pp. 1–7, doi: 10.1109/ICOMICON.2017.8279028.

29. J. Xiong, Z. Yao, J. Ma, X. Liu, Q. Li, and T. Zhang, "PRAM: Privacy preserving access management scheme in cloud services," in *Cloud Computing 2013- Proceedings of the 2013 International Workshop on Security in Cloud Computing*, 2013, pp. 41–46, doi: 10.1145/2484402.2484412.

30. K. W. Nafi, T. S. Kar, S. A. Hoque, and M. M. A. Hashem, "A newer user authentication, file encryption and distributed server based cloud computing security architecture," *International Journal of Advanced Computer Science and Applications*, vol. 3, no. 10, pp. 181–186, 2013, Accessed: March 17, 2021. Online. Available: http://arxiv.org/abs/1303.0598.

31. A. K. Marnerides, M. R. Watson, N. Shirazi, A. Mauthe, and D. Hutchison, "Malware analysis in cloud computing: Network and system characteristics," in *2013 IEEE Globecom Workshops, GC Wkshps 2013*, 2013, pp. 482–487, doi: 10.1109/GLOCOMW.2013.6825034.

32. A. Bouayad, A. Blilat, N. E. H. Mejhed, and M. El Ghazi, "Cloud computing: Security challenges," in *CiSt 2012- Proceedings: 2012 Colloquium in Information Science and Technology*, 2012, pp. 26–31, doi: 10.1109/CIST.2012.6388058.

33. E. Simmon and R. Bohn, "An overview of the NIST cloud computing program and reference architecture," *Advanced Concurrent Engineering*, pp. 1119–1129, 2013. doi: 10.1007/978-1-4471-4426-7_94.

34. J. Steinke, B. Bolunmez, L. Fletcher, V. Wang, A.J. Tomassetti, K.M. Repchick, S.J. Zaccaro, R.S. Dalal, and L.E. Tetrick, "Improving cybersecurity incident response team effectiveness using teams-based research," *IEEE Security & Privacy*, vol. 13, no. 4, pp. 20–29, 2015.

35. A.A. Ramaki and R.E. Atani, "A survey of IT early warning systems: Architectures, challenges, and solutions," *Security Communication Networks*, vol. 9, no. 17, pp. 4751–4776, 2016, doi: 10.1002/sec.1647.

36. A. Ahmad, S. B. Maynard, and G. Shanks, "A case analysis of information systems and security incident responses," *International Journal of Information Management*, vol. 35, no. 6, pp. 717–723, 2015, doi: 10.1016/j.ijinfomgt.2015.08.001.

37. R. Ruefle, A. Dorofee, D. Mundie, A. D. Householder, M. Murray, and S. J. Perl, "Computer security incident response team development and evolution," *IEEE Security & Privacy*, vol. 12, no. 5, pp. 16–26, 2014, doi: 10.1109/MSP.2014.89.

38. N. H. Ab Rahman, N. D. W. Cahyani, and K.-K. R. Choo, "Cloud incident handling and forensic-by-design: Cloud storage as a case study," *Concurrency and Computation: Practice and Experience*, vol. 29, no. 14, p. e3868, 2017, doi: 10.1002/cpe.3868.

39. S. H. Albakri, B. Shanmugam, G. N. Samy, N. B. Idris, and A. Ahmed, "Security risk assessment framework for cloud computing environments," *Security Communication Networks*, vol. 7, no. 11, pp. 2114–2124, 2014, doi: 10.1002/sec.923.

40. A. Murray, G. Begna, E. Nwafor, J. Blackstone, and W. Patterson, "Cloud service security & application vulnerability," in *Conference Proceedings - IEEE SOUTHEASTCON*, 2015, vol. 2015, doi: 10.1109/SECON.2015.7132979.

41. A. G. Kumbhare, Y. Simmhan, M. Frincu, and V. K. Prasanna, "Reactive resource provisioning heuristics for dynamic dataflows on cloud infrastructure," *IEEE Transactions on Cloud Computing*, vol. 3, no. 2, pp. 105–118, 2015, doi: 10.1109/TCC.2015.2394316.
42. S. Narula, A. Jain, and Prachi, "Cloud computing security: Amazon web service," in *International Conference on Advanced Computing and Communication Technologies, ACCT*, 2015, vol. 2015, pp. 501–505, doi: 10.1109/ACCT.2015.20.
43. M. Soltys, "Cybersecurity in the AWS cloud," arXiv, 2020, Accessed: March 18, 2021. Online. Available: http://arxiv.org/abs/2003.12905.
44. M. Copeland, J. Soh, A. Puca, M. Manning, and D. Gollob, Microsoft azure and cloud computing. In *Microsoft Azure,* pp. 3–26. Apress: Berkeley, CA, 2015.
45. N. J. Mitchell and K. Zunnurhain, "Google cloud platform security," in *Proceedings of the 4th ACM/IEEE Symposium on Edge Computing, SEC 2019*, 2019, pp. 319–322, doi: 10.1145/3318216.3363371.

12 The Cyber Artificial Intelligence Platform for Cloud Security

P. Manju Bala, S. Usharani, R. Rajmohan,
S. Jayalakshmi, and P. Divya
IFET College of Engineering

CONTENTS

DOI: 10.1201/9781003219880-12

12.1 INTRODUCTION

The cloud is one of the systems for storing massive amount of data, and it provides storage and computing power. This is a location that is necessary for many businesses and organizations to protect their information, and there is a risk of being compromised by others. The cyber AI framework is proposed for this cloud protection to secure the cloud. Intrusion, illegal users, vulnerability, protocols or functionalities, data theft, lack of accessibility, remote information sharing, suspicious members, cyber assault, and DoS attacks are several cloud problems. Generally, the area of AI is where computers may spontaneously perform tasks usually requiring human intelligence. It includes artificial intelligence and machine learning used to improve knowledge and develop skills. Machine learning is the subset of DL where it gains knowledge through algorithms used by human interaction, artificial neural network, and study from vast volumes of data.

Cyber AI is an inner understanding that investigates knowledge and behaviour, like the human immunity system. It includes creating millions of predictions for emerging facts depending on uncertainty. Cyber AI researchers simultaneously sort, observe, and update on the complete spectrum of complex employee security breaches. Through this, the researcher will secure the protection of the cloud. One type of cloud database framework involved in using the cloud to provide a full range of services is cloud computing. Systems may be in the form of basic programs built for the purpose of performing a particular operation, or they may be an interface shared over the cloud, or some network related to the software shared over the

Internet. Cloud computing platforms can be classified into three classes, namely software as a service (SaaS), infrastructure as a service (IaaS), and platform as a service (PaaS), depending on the above description. Some specific elements characterized earlier cloud storage. Service on request is an essential aspect of cloud services that indicates the customer may request the appropriate number of services needed to fulfil the requirements of the customer. In terms of SaaS, PaaS, and IaaS, in compliance with the customer demand, the service offers services such as applications, platforms, and infrastructural facilities. In all facilities, the business information of the customer is submitted to provide personal details and customer rules. It is important to safeguard confidential details and industry guidelines and enhance service quality. In SaaS, PaaS, and IaaS, it is therefore necessary to provide protection. The level of protection is very limited, not just for facilities, in spite of fraudulent users, access authorization, connection or path, and data processing. Some other remarkable features of cloud computing are that it is extremely flexible in nature and it allows companies the opportunity to build up or level down demand-based services, resulting in major cost savings. An important feature found in cloud computing is that the services are not firmly attached to the customer; instead, the services are calculated in terms of their use when contributing to the cloud provider at a highly detailed level and are paid depending on their use. This principle of pay per use found in cloud computing has greatly reduced the expense spent in providing services for fixed service-oriented applications and services. Such features have seen a significant change in the usage of cloud-based platforms by various industries. Models for the application of cloud computing include the use of personal cloud solutions such as OpenStack and VMWare that respond to personal user demands, public cloud solutions such as AWS, Google Cloud services, and Microsoft Azure, and a mix of multiple cloud frameworks called private and public cloud systems. Nevertheless, it provides services to a broad community base as public cloud services have been used, and because of that, the public cloud framework is of a low- and mid-type. The low- and mid-design of the cloud system encourages knowledge exchange, which in turn raises the risk of sharing certain data and resources from other users. So many global companies and sectors are also reluctant to use cloud environments only for the reason that by using cloud environments, business-sensitive details are not exchanged or breached.

The complexity and relevance of cloud security and the necessity for specialized and innovative methods to facilitate security in the cloud system have been highlighted in this perspective. Fortunately, in the promotion of cloud protection, many cloud-based data authentication specifications and guidelines introduced in recent decades have played a crucial role. In addition, multiple access control strategies and sophisticated identity identification, mitigation, and surveillance tools have facilitated improved cloud environment safety policies.

12.2   ISSUES THAT OCCUR IN THE CLOUD SECURITY

Ninety-four percentage of associations are tolerable to very worrisome about cloud security. At the point when some information is obtained about what are the greatest security dangers confronting public mists, associations positioned misconfiguration

(68%) most noteworthy, followed by unapproved access (58%), uncertain interfaces (52%), and seizing of records. Today's market has security risks and concerns regarding the highest level cloud. Almost every organization has incorporated distributed computing to varying degrees within their operations. Nonetheless, with this cloud adoption come the requirements to ensure that the organization's cloud security strategy can mitigate the top cloud security threats.

12.2.1 ILLEGAL ACCESS

Cloud-based solutions are available directly from the open network and are not restricted to the organization's perimeter. Although this can help the administration use the system, it also makes it easier for an attacker to access unwanted cloud-based products. Incorrectly built security or established credentials allow an adversary to connect an organization's data without the organization's knowledge.

12.2.2 HIJACKING OF ACCOUNT

Many people have a minimal capability of concealing their secret words, which consists of recurring secret words and weak passwords. Inaccurate implementation of the cryptographic algorithm increases the likelihood of attacks by malware emails and data breaches by allowing a single stolen secret key to be used on multiple entries. Records requisitioning is one of the more severe cloud security threats, as organizations rely more and more on cloud-based applications and infrastructure for core business functions. When credentials are stolen, both clients and systems with administrative access can use the stolen credentials to control Internet identity. In addition, organizations that use the cloud typically cannot detect and respond to threats as they could if they were running in house.

12.2.3 EXTERNAL DATA SHARING

The cloud is built to facilitate the sharing of data in an uncomplicated manner. Many websites send an email thanking the person or letting anyone with the URL to the shared resource access it. While easy information exchange is beneficial, obtaining it is not always easy. As a threat, it is a significant cloud potential. Shared resources, such as training sessions, can be restricted because they are easier to share using connection-based communication than explicitly inviting each training partner. When one person has a shared connection, that connection can be provided to another person, whom a hacker can then hack into the shared resource. Distributing connections in this manner generally prevents only one recipient from being denied access to the network.

12.2.4 CYBERATTACK

In enterprises, security threats are targeted, which influences which computers will be targeted. Most cloud-based systems are available on the open network, are frequently inaccurate, and store important data. It is also possible that frequent attacks could be made against the cloud, increasing the chances of success. Hackers commonly target cloud-based associations, especially since the security problems of cloud computing became widespread.

12.2.5 DoS Attack

For many organizations, collaboration is easier and more effective, thanks to the cloud. To hold business-critical data, the cloud is used for operating internal and client-facing services. DoS attacks, which specifically target cloud infrastructure, are hazardous for organizations due to the wide ranging impacts. This, coupled with extortion attempts by assailants who employ distributed denial-of-service (DDoS) attacks, makes them a serious threat to a business's cloud computing-based resources.

12.3 CYBER ARTIFICIAL INTELLIGENCE ANALYSIS

The following are the different information sources from where information is gathered and afterwards examined.

12.3.1 Customer Information

Customer information will be collecting and examining client access and exercises from AD, proxy, VPN, and applications.

12.3.2 Submission Information

Submission information includes the collection and examination of calls, information trade, and orders along with the web application firewall information for introducing the specialists on the application.

12.3.3 Resultant Data

Resultant data is used to analyse the inner endpoints, for example records, measures, memory, library, associations, etc. are obtained by introducing specialists.

12.3.4 Web Data

Web data comprise of network forensics and analytics products collecting and dissecting the bundles, net streams, Domain Name System, and intrusion prevention system information by introducing the organization apparatus.

12.4 THE IMPACTS OF AI ON CYBER SECURITY

It may involve two strategies to define artificial intelligence. To begin with, it is a methodology that aims to discover the basis of intelligence and create autonomous systems by incorporating data, logic, perception, and the willingness to make gadgets intelligent. To put it another way, people make gadgets using their intelligence. This intelligence may read, analyse, determine, and perform problem solving as a human intelligence. Researchers, however, describe artificial intelligence as a discipline that investigates and establishes techniques to resolve issues of uncertainty that cannot be resolved without knowledge being adopted. For example, depending on huge quantities of data, researchers can create an AI framework for decision-making and real-time analysis. Artificial intelligence has contributed to developments in several scientific fields and technological fields in recent decades, such as computerized robots, facial recognition, processing of linguistic structure, and intelligent agents.

The rapid growth of computing technologies and the cloud is creating a significant impact on people's daily work and interests. Furthermore, it has also generated a slew of cyber security threats: First, the accumulation of data makes a thorough review obsolete. Second, dangers are increasing at a fast pace, resulting in the emergence of new, superficial environments and quickly adaptive threats. Additionally, hazards have an impact on various distribution, duplication, and mitigation approaches, making them challenging to recognize and analyse. Furthermore, the cost of eliminating risks must be considered. It takes a lot of time, energy, and effort to create and implement algorithms. As a result, finding and training experts in this field is difficult and expensive. This is the case the majority of the time; numerous variations in danger occur and spread on a regular basis. Some cyber security concerns will also necessitate AI-based solutions.

12.4.1   THE POSITIVE USES OF AI

Artificial intelligence is now being used in the area of cyber security to enhance defensive skills. AI will be able to analyse vast volumes of information with effectiveness, precision, and volume, depending on its efficient optimization and data processing technologies. In order to detect related threats in the context, even if their trends shift, an AI framework should take benefit of whatever it observes and recognize previous threats. Undoubtedly, when it refers to crime defence, AI has some strength in the following facets.

Modern and complicated improvements in attack versatility can be discovered by AI: Traditional intelligence focuses on the experience and depends highly on proven hackers and threats, providing space for blind spots in different threats when predicting abnormal activities. By intelligent technology, the shortcomings of old defence technology are now being discussed. Wealthy intranet behaviour, for example, can be tracked, and therefore any major variation in user authentication activities can signify a possible existential threat. The computer would improve the validity of the acts if the identification is accurate and becomes more susceptible to identifying similar models in the context. The computer can understand and develop better to detect anomalous, quicker, and more precise activities with a greater volume of information and many scenarios. This is particularly important as cyberattacks are being more advanced, and attackers are developing different and inventive techniques.

The amount of information can be managed by AI, by designing intelligent protection technologies to identify threats and react to threats. AI can boost security measures. For security communities, the number of safety warnings that occur regularly can be quite daunting. Detecting and responding to assaults interactively has reduced the job of cyber security specialists and can assist spot assaults more quickly than other methods. Whereas a large amount of safe data is produced and delivered over the Internet each day, information security experts will have growing difficulty tracking and identifying threat factors quickly and precisely. Artificial intelligence will aid in the detection and surveillance of hazardous behaviours. This will aid cyber security personnel in adapting to situations they have never encountered before, as well as eliminate time-consuming people investigation.

Over period, an AI protection framework can evolve to react faster to attacks: AI tries to identify attacks depending on user actions and the operation of an entire system. Over period, the AI safety system knows about the daily activity and behaviour of the system and offers a background of what is usual. From there, to identify intrusions, any anomalies from the standard can be identified. AI algorithms tend to be an optional field of analysis that strengthens cyber security protection controls.

Most AI techniques, such as cognitive computing, machine learning, expert systems, automated immune systems, data analysis, problem solving, heuristics-based systems, deep learning, and machine learning, are being used to counter attacks. Among these strategies, however, deep learning and machine learning have currently gained a great deal of publicity and have earned the most accomplishments in the fight towards cyberattacks.

12.4.2 DRAWBACKS AND LIMITATIONS OF USING AI

Datasets: Designing an artificial intelligence framework requires a large amount of input samples, and it can require a significant period and amount of money to collect and analyse the sample data.

Resource specifications: An enormous number of resources, especially storage, information, and computational power, are needed to construct and retain the basic system. In addition, the professional personnel required to incorporate this technology entail considerable costs.

Fake notifications: For customers, repeated fake notifications are a concern, affecting enterprise by possibly frustrating any reaction that is required and consequently reducing performance. The perfect procedure is an exchange between reducing fake notifications and retaining the level of protection.

AI-based device attacks: Hackers can use different intrusion methods that involve AI systems, such as confrontational inputs, data manipulation, and software theft. The illegal use of AI is one significant factor to be taken into consideration. This framework can also be used as a method to increase risks. Cybercriminals, for instance, can exploit the ML techniques to produce a system malicious model that is difficult to detect. Moreover, AI will be able to further personalize the malware system and increase the scale of the intrusion, allowing the threat more possible to be successful.

12.5 AI METHODOLOGY FOR CYBER SECURITY

12.5.1 LEARNING METHODS

AI is an information technology field that aims to create a modern form of smart algorithm that reacts like human intelligence. Systems need to train to attain that objective. To be more specific, using the ML techniques we have to educate the machine. Typically, learning methods help to increase efficiency by understanding and practising from knowledge in achieving a mission. At present, there are three main categories of learning methods used to educate devices:

Supervised learning: Supervised learning involves a method of learning with a broad and recognizable collection of already labelled data. As a selection process or correlation process, these learning methods are also used.

Unsupervised learning: Unsupervised learning methods use unidentifiable datasets, as compared to supervised learning. These methods are also used for data clustering, dimensionality reduction, or depth estimation.

Reinforcement learning: It is a kind of learning method that is focused on encouragement or penalties to determine the right acts. In circumstances where data are minimal or not provided, reinforcement learning is effective.

12.5.2   MACHINE LEARNING METHOD

Machine learning is a category of AI that seeks to motivate processes without having specifically configured by using knowledge to understand and develop. ML has broad links to computational approaches that facilitate knowledge extraction, pattern discovery, and information forming assumptions to be processed. The ML model has various forms, and they can typically be grouped into the following key areas: reinforcement learning, supervised learning, and unsupervised learning. In the information security domain, the standard ML algorithms are support vector machines, decision trees, Bayesian classification, association rule mining, etc.

12.5.3   DEEP LEARNING METHOD

Deep learning is a discipline of machine learning, and it requires knowledge to educate machines how only people are trained to do stuff at that period. Its purpose resides in the human mind and brain cells' functioning methods to interpret signs. The essence of deep learning is that its efficiency begins to improve as we build more comprehensive NN and educate them with the data as necessary. Its excellent output in complex data is the main significant benefit of DL over traditional ML. DL approaches can include reinforcement learning, supervised learning, and unsupervised learning, equivalent to machine learning algorithms. DL's advantage is the advantage of unsupervised learning to randomly pick functions. In the cyber safety realm, the typical deep learning algorithms widely used are recurrent neural network, gated recurrent unit, convolutional neural network, stacked RNN, deep belief network, etc.

12.5.4   BIO-INSPIRED METHOD

The bio-inspired computation is a subset of artificial intelligence that has evolved as among the major learning methods in the recent decades. In order to overcome a wide variety of difficult intellectual and specific field issues, it is a set of intellectual models and techniques that follow bio-inspired behaviours and traits. The following approaches are most widely used in the field of cyber security by many bio-inspired methods: artificial immune system, genetic algorithm, ant colony algorithm, etc.

12.6 AI TECHNIQUES FOR PRESERVING AGAINST CYBERSPACE ATTACKS

In recent times, researchers have suggested various approaches that use AI approaches to recognize or categorize ransomware; identify system breaches, spoofing, and malicious attacks; address advanced cyber threat (ACT); and recognize domain created by algorithms developed by domain generation algorithms (DGAs). There are four main techniques used against cyberspace attacks such as malicious identification, intrusion detection system, SPAM, and phishing identification, which undermine the disputing of ACT and the identification of DGA. The main AI techniques for cyber security are outlined in Figure 12.1.

12.6.1 IDENTIFICATION OF MALWARE

For several forms of malicious software, malware is a common word, such as bugs, zombies, computer viruses, hacks, adenoviruses, and today, a common cyberattack technique is malware. The effect of malware on the digital society is immense; so a significant number of analyses have been conducted to avoid and reduce malware by implementing AI techniques. For malware identification and protection, the most current and notable developments use knowledge. In Ref. [9], the authors introduced machine learning to build a digital platform focused on virtual memory accessing characteristics for hardware-assisted detection of malware. Logistic regression, a SVM, and a RFC were used in the proposed process and executed on the RIPE comparison set for the simulations. The authors stated that with a <5% false-predictive rate, the system has a real predictive accuracy of 99%. After that, a method for identifying and finding malicious code including knowledge discovery and ML specification was proposed by the authors in Ref. [10]. In this paper, for identification, both stamp-based and outlier-based features were analysed. Investigational findings

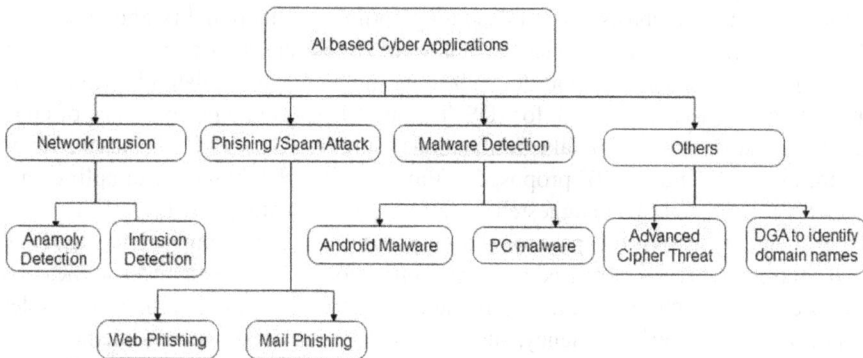

FIGURE 12.1 AI techniques for cyber applications.

revealed that certain related approaches were surpassed by the suggested approach. The operational code was shown as a chart and incorporated in the Euclidean space; then, each variable was classified as malware or neutral by one classification model or an aggregate of classification models. The experiential output indicated that the suggested method is effective with a little fake notifications rate and higher discovery rate. For sophisticated malware identification, in Ref. [12], the authors developed a deep learning framework. In this work, to identify unidentified malware, they used an AE stacked up with layered restricted Boltzmann machines (RBMs). The author argued that, relative to conventional superficial learning approaches and deep learning approaches, clustered deep learning structures may increase the cumulative results in malware identification. A recent trend in malware analysis testing has concentrated on web adware in general and robot adware in particular. A big development in this field was ML with DL. In Ref. [13], to classify malware, a DCNN (deep convolutional neural network) was introduced. The series of raw OpCode from a reassembled program has been used to recognize malware. A SVM and the most important credentials from all authorization data were used by the authors in Ref. [14] to differentiate among positive and harmful apps. The authors introduced unique ML algorithms for malware identification in Ref. [15], especially rotating forest. To identify application malware, an ANN and the actual sequence of application programing interface process requests were used in Ref. [16]. In order to improve the reliability and performance of large-scale software malware identification, a training method in Ref. [17] implemented a dual method depending upon autoencoder and CNN. The use of bio-inspired approaches for the identification of malware was also other experimental paths that captured scientists' interest. These approaches were used primarily for feature enhancement and classifier parameters enhancement.

12.6.2 DETECTION OF INTRUSION

Intrusion detection is a method designed to safeguard the system against potential attacks, breaches, or immediate attacks. Because of their versatility, versatility, efficient measurements, and fast learning, AI-based methods are ideal for trying to develop IDS and accomplish other approaches. Therefore, many scientists have been studying smart approaches to enhance IDS efficiency. In order to decrease fake information, the emphasis was on designing optimized functionality and enhancing the classification model. Some significant example findings are described as follows.

Support vector machine and deep learning methods with updated k-means were integrated [20] as a framework for IDS. The model reported a performance of up to 95.75% accuracy and 1.87% fake notifications using the KDD'99 Cup datasets. In the meantime, the author in [21] proposed a framework for IDS based on sampling with a least square SVM. The suggested approach was evaluated, and it is achieved by experimental setup with respect to reliability and effectiveness through the KDD'99 Cup datasets. A semi-supervised learning method for IDS based on uncertainties was proposed by the authors in Ref. [22]. In their work, they used unidentifiable examples to boost the classifier's efficiency, supported by a supervised learning method. The algorithm was evaluated, and it surpassed other comparable methods on the KDD'99 Cup datasets.

12.6.3 Masquerading and Malware Identification

A masquerading attack is a cyber-attempt to steal a person's personal or financial details. Masquerading attacks are presently one of the most dangerous cyber risks. Many new advanced strategies have been used to address these issues. The authors introduced a fraud detection system called Spoofing Mail Protection system (SMPS) in Ref. [23], which combined the growing computational model and classification algorithms. The accuracy level of their model was 98.7%, with a false correctness rate of 1.9%. The researchers designed an anti-phishing strategy in Ref. [24], which used a variety of machine learning methods and 19 attributes to distinguish spammers from legitimate users. The authors reported that a 99.38% true-positive rate was reached by their methodology. Another method that is combined with the Monte Carlo equation and risk reduction theory which is implemented by neural network to classify phishing websites. The experimental findings showed that the model achieved an effective identification rate of 97.71% and a fake identification rate of 1.7%. Another study implemented a real-time anti-phishing framework that used seven dissimilar classification methods and functionality based on natural language. With a 97.98% accuracy rate, the method obtained positive results, as per the authors. A further study developed a stacked model using hyperlink and markup functionality to identify phishing web pages by integrating LightGBM, XGBoost, and GBDT. The authors recorded that their method achieved a precision rate of 98.60%. The word 'spam' applies to unwanted spam emails. Security problems and improper information may result from spam emails. In order to resolve the disadvantages of these cyber threats, researchers have newly implemented various groundbreaking, smart approaches to create spam filter mechanisms.

12.7 ISSUES OF ARTIFICIAL INTELLIGENCE PLATFORM FOR CLOUD SECURITY

The emergence of intelligent assistants to boost the efficiency and reliability of cyber security data analysis is particularly embraced by several implementations of AI in cyber security. In cloud computing, open stack frameworks used to proceed; the lack of privacy knowledge seems to be a perfect solution for the existence of strong algorithms and machine learning applications. Fortunately, the reliability of such technologies in information and design faces specific issues [25].

A well-known reality in the security industry is expertise inadequacy. Hiring and recruiting qualified security experts is challenging, with 52% of firms reporting that in 2018 they had a challenging deficiency of cyber security expertise. The wealth of emerging technology, warnings, and data continues to grow, placing increased responsibility and demand on security teams to understand and protect emerging developments. The number of fresh flaws found in 2017, which more than doubled from 2016, is a real illustration. The researchers of cyber security is to accept AI is predictable. Security teams will be well prepared with measurable knowledge and concentrate on mitigating crucial security issues by allocating the first stage of analysis and description to the 'bots' completely or partially, thereby enhancing the efficacy of threat detection and response:

- Sorting warnings
- Identifying the most critical issues
- Taking precedence assignments as per the threat
- With security threats, tracking activities and methodologies
- Detecting suspicious behaviours.

12.7.1 CHALLENGES IN AND SOLUTIONS TO DATA SECURITY

The next obstacle for businesses, beyond the AI excitement, is to consider what the security best practices are for each latest AI system. It comes up with attack surfaces (and security flaws) that attackers can use to recover data, as with any emerging technologies:

A. **Platform:** Many businesses use cloud services, e.g. Azure, but this does not eliminate the requirements to monitor and implement best practices for cloud security configuration implementation, even though we respect our cloud services.

B. **Virtual cloud:** Keeping track of all properties, such as databases, networks, and memory, storage, is important for infrastructure and services.

C. **Security of applications, APIs, and staffing levels:** Often the most noticeable and open aspect that can be abused by basic web assaults. They can also be available publicly!

D. **Protection of messages:** Communication system must adopt standards of privacy and honesty.

E. **Capacity:** With network access, privacy, honesty, and accessibility, the stored data must be secured.

F. **Data security during implementation:** A more complicated topic than the others, with significant development in the field of asymmetric cryptography.

An approach to substances D and E is encryption. For instance, even though people are utilizing a cloud service, they can retain the authority of authorization codes and use cryptographic algorithms. Both cloud services provide authentication techniques so that, while retaining hold of the codes, users can put in place sufficient safeguards. New studies on homomorphism encryption by enabling sophisticated data analytics to be carried out in encrypted data without sacrificing authentication are anticipated to provide a significant role in cloud data services for attack substance F.

Any individual with a corporate credit card can turn up latest cloud services, unlike conventional systems; financing and expenditure reviews are one of the ways to reclaim the track of virtual cloud (B). Even if the identification is often interrupted because it gets involved in the invoice date, the on-demand payment helps to identify new requirements and enable the resources seemed to be monitored more accurately. A safer practice in using the cloud will be to set up more intermittent, even constant, tracking of the use of properties. This allows the identification of irregularities and the costs to be more comprehensive. Again, APIs that provide us with automatic demand forecasts (and payment) make this control feasible.

For substance A, the configuration benchmarks are good places to start. There is one for applying best security measures for Amazon Web Services and another for Microsoft Azure. Organizational and automatic monitors cover the 'standards'. More sophisticated measures, such as those drafted by the cloud protection association, will supplement them.

12.8   CLOUD SECURITY

In order to enhance the reliability of a cloud network, cloud security comprises all software technologies and organizational strategies and measures used. Cloud protection is intended to assure the data privacy, and also the structure of the cloud themselves. This will depend on the requirements, the cloud service that is used, and the strategies they have for cloud protection. Notice that, the business person and service provider, that use the cloud protection processes to secure the data is a shared obligation between authenticated users.

The key significant cloud security goals are described below:
- Protect your records, including any property rights, in the cloud (IP)
- Preserve the confidentiality of the employees and consumers and private identifying data
- Have a mechanism to maintain adherence to regulatory protocols
- Create rules to validate cloud network access devices and users.

It is possible to customize cloud privacy to the specific needs of the company. The endless complexities of verifying users and limiting traffic from sources that are known to be harmful will cover these needs.

12.9   STRATEGIES FOR CYBER SECURITY EFFICIENCY METRICS

12.9.1   INSUFFICIENT ELIMINATION OF DATA

The collection of information about occasions that are not useful for the recognition period is taken as excess data. In this way, data are collected with the aim of extending the presentation. As shown in Figure 12.2, the data are directed to the programming advancement section after the removal of insufficient information to classify digital breaches. Finally, using perception components, the results are predicted.

12.9.2   FEATURE EXTRACTION AND SELECTION

Feature extraction and selection is shown in Figure 12.3. The component extraction and highlighted choice cycles permit equal handling capacities to speed up the determination and extraction measure. At that point, the removed element dataset is sent onto the information examination module that plays out an alternate activity to break down the abatement in the size of the dataset to distinguish digital breaches.

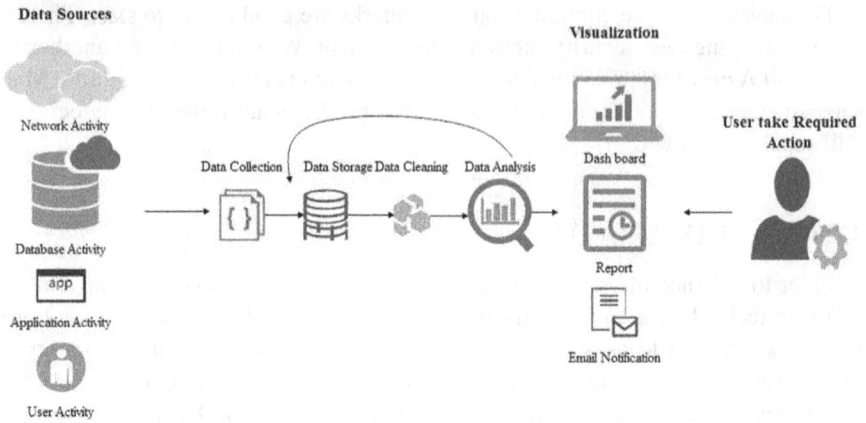

FIGURE 12.2   Insufficient elimination of data.

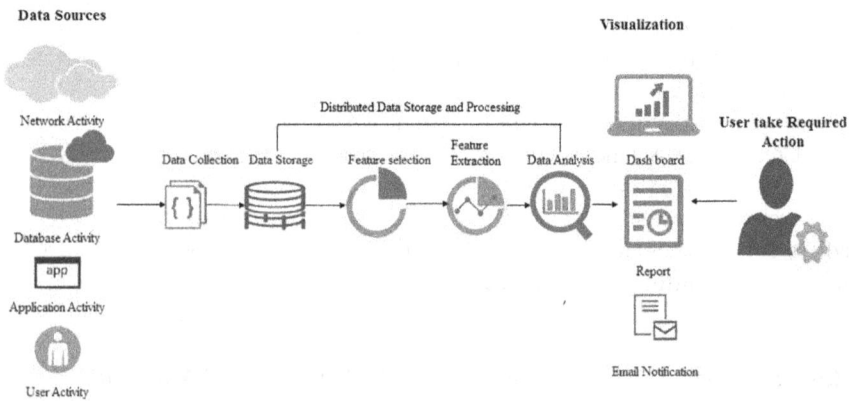

FIGURE 12.3   Feature extraction and selection.

Precautionary measures are recommended in the situations of an attack that can be interpreted by the client using the interpretation section. A business or customer may find a way to minimize or thwart the effects of the attack if these attack warnings go under notifications.

12.9.3 DATA CUT-OFF

The information cut-off segment forces the cut-off by dismissing security occasions that arise after the association of an organization or cycle has arrived at its generally characterized limit. Any security occasion that arises after as far as possible doesn't contribute without a doubt to the assault discovery measure, in this manner, dissecting these sorts of security occasions infers an additional weight on information

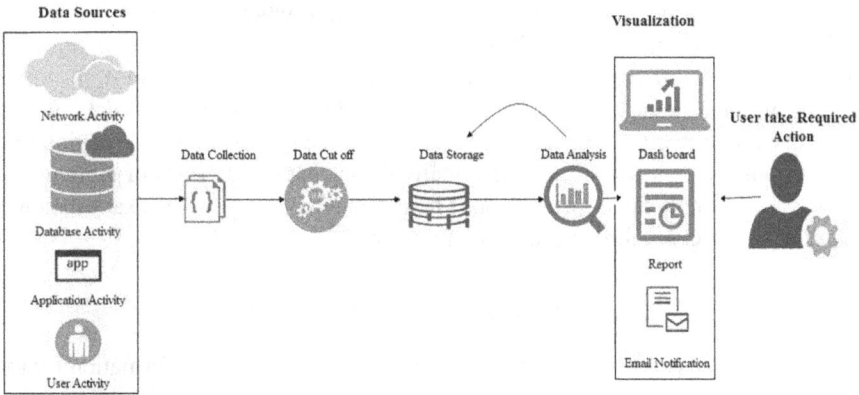

FIGURE 12.4 Data cut-off.

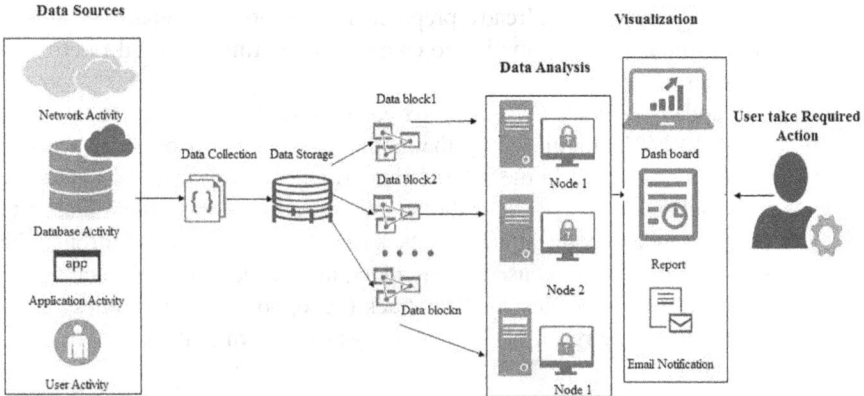

FIGURE 12.5 Parallel processing.

handling assets with unmistakable growth. The information stockpiling substance can store the security occasion information left after cut-off as shown in Figure 12.4. The information investigation module reads the put away information to dissect it for identifying digital assaults. Eventually, the after-effects of the examination are imaged to a client through a representation substance, which permits a client to make the imperative move upon the appearance of each extraordinary alarm.

12.9.4 PARALLEL PROCESSING

The information authority substance catches security occasion information from various assets relying upon the various sorts of security investigation and security necessities of a particular undertaking that is shown in Figure 12.5. The information gathered conveys the caught information to an information stockpiling substance, which stores the information. There are numerous approaches to store information, for example Hadoop File System, HBase, and relational database. To apply equal handling, the put away information should be disseminated into data blocks. Subsequent to dividing, information is imported in the information examination segment through various hubs working equally dependent on the rules of a circulated system, for example Spark or Hadoop. The outcome obtained by the examination is imparted to the client through the perception part.

12.9.5 TRAINING MODELS USED FOR CYBER AI

The information assortment substance catches security occasion information for the preparation cycle of a security investigation framework. The preparation information can be snatched from sources inside the venture where a request should be conveyed. Subsequently the gathering of the information for preparation, the information readiness segment begins the way towards setting up the information for model preparation by applying different channels. From the point forward, the chose ML calculation is executed in the already prepared information to prepare an assault identification method. From estimation to estimation, the time required to create a design (i.e. processing period) varies.

After the model in Figure 12.6 is prepared, it is supposed to investigate how digital attacks can be distinguished by the model. Information is obtained with an activity for model research. Via the information arrangement system, the data to be attempted are isolated and taken into the attack position system, which is used to examine the data to identify the attacks on the basis of the criteria found through the planning phase. The duration is used by an attack identification model to close if a specific flow of details associates with an attack (i.e. option time) and relies on the quantified evaluation. The result of the data analysis is shown to the customer via a perception variable.

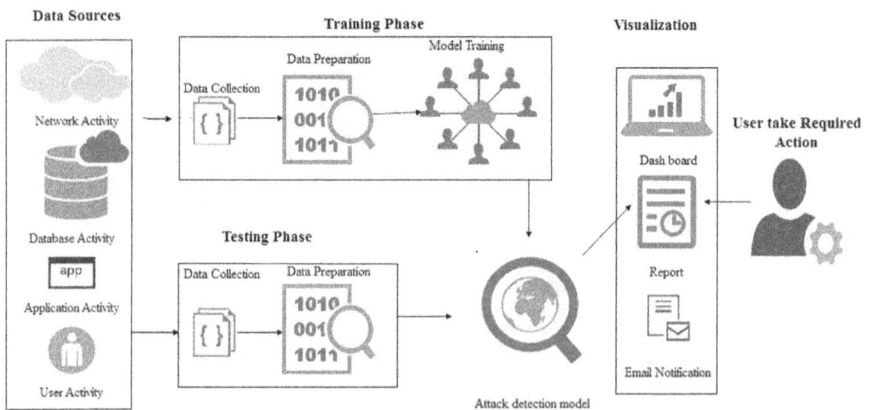

FIGURE 12.6   Training models used for cyber AI.

12.9.6 ACCURACY IN SECURITY MODELS

12.9.6.1 Alert Correlation

Afterwards, the information aggregation section grabs security opportunity details from different assets; collected data are then stored in the supplying of data and redundant to the per-processor section for the application of pre-handling techniques on the raw data. In the ready review module, which conducts research on the data for the identification of attacks, the pre-treated data are swallowed. In Figure 12.7, it is necessary to remember here that the alert analysis unit articulates the data in an outdated style either depending on inconsistency or using abuse-based analysis or both. The alarms generated will be sent to the ready validation unit, which uses different processes to decide if an alarm is dishonestly safe. At this stage, the warnings classified as false positives are rejected.

The splendid and very much orchestrated cautions are then sent to the ready connection module for additional investigation. From that point forward, the cautions are connected (i.e. intelligently connected) utilizing various strategies and calculations, for example rule-based relationship, situation-based connection, fleeting relationship, and measurable connection. In order to collect the requisite logical warning data, the alert connection unit communicates with the data stockpiling. Through the representation unit, the implications of the interaction are published. At last, neither a robotized reaction is created, nor a security chairman plays out the examination of the danger and reacts likewise.

12.9.6.2 Signature-Based Anomaly Detection

In Figure 12.8, the information assortment part gathers security-applicable information from various assets. From that point onwards, the gathered information is put away by the information stockpiling module. Next, the information is brought into the mark-based recognition part that plays out the investigation on the information

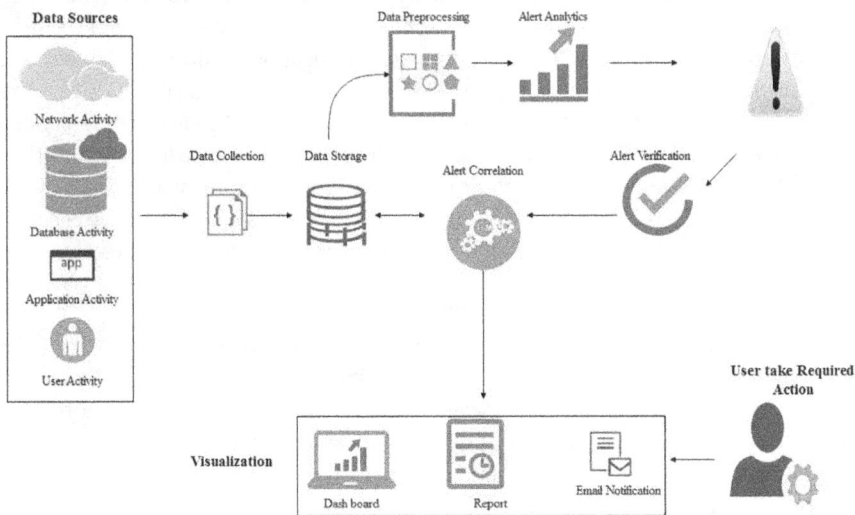

FIGURE 12.7 Alert correlation.

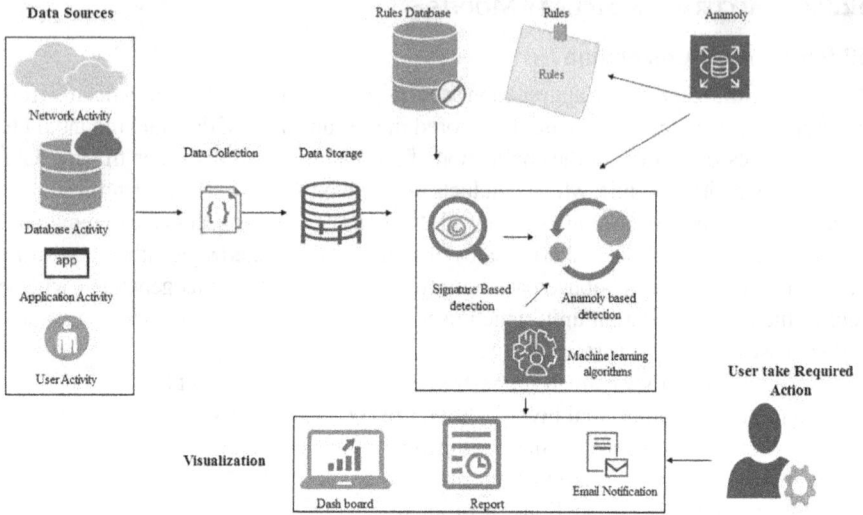

FIGURE 12.8 Signature-based anomaly detection.

to identify examples of the assault. For such investigation, this part gives the benefit of the pre-planned standards from the information base of the states that recognize examples of the assault. In the event that any match is recognized, an alarm is straightforwardly created through a representation module.

In the event that the mark-based identification part doesn't distinguish any example of assault in the information, the information is passed to the peculiarity-based location segment for identifying obscure assaults that can't be recognized by the mark-based discovery segment. A peculiarity is characterized as the abnormal conduct, or example of the information. This specifically shows the presence of the mistake in the framework. The peculiarity-based identification module investigates the information utilizing calculations of AI to distinguish deviations from typical conduct. At the point when an irregular (deviation) is recognized, an alarm is created through the representation module. At a similar example of time, the peculiarity is characterized as an assault example or rule and sent to the information base of the standards. Utilizing thusly, the principles information base is ceaselessly refreshed to empower the mark-based location segment to distinguish an assortment of assaults.

12.9.6.3 Attack Detection Algorithm

The information assortment unit gets security occasion information for preparing the security scientific framework for identifying digital assaults. The preparation information can be gathered from various assets inside an undertaking where a request should be conveyed. After the cycle of information assortment identified the preparation information, the information planning unit readies the information for preparing the model by inferring various channels and methods of highlight extraction.

First, the ready planning data enact the preparation of the module for attack detection shown in Figure 12.9. When the unit is ready, it is authorized to investigate

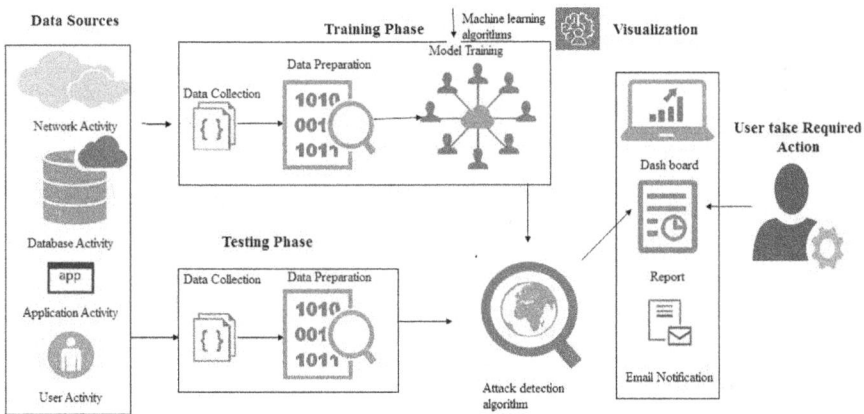

FIGURE 12.9 Attack detection algorithm.

whether digital attacks can be recognized by the system. The data are obtained from a firm to approve the method. The sample data are ready to be submitted into the system for attack identification. The ready test data are introduced into the attack position model, which conducts the analysis on the basis of the criteria collected during the planning phase. The malicious or legitimate interruptions are called in the introduced test knowledge instances. Via the perception unit, the investigation findings are imaged to a client. At the time of malicious or attack instances, a client may make fast appropriate movements that involve closing a few ports or cutting off the firm's affected parts to avoid further destruction.

12.10 GPU-BASED CNN-MSVM CLOUD SECURITY SYSTEM

The aim of the CNN-MSVM cloud protection framework based on GPU is to develop a reliable and strong cloud security relying on machine learning by analysing data trends and identifying deviations and irregularities found using these ML approaches in the cloud setting. Detecting abnormalities with a high rate of reliability and clarity is the primary reason behind using a ML-based approach to support cloud protection. The dataset is examined by checking the transfer rate of the data, the time period during the contact, and other variables in the dataset to classify the attacks. By calculating the TP, TN, FP, and FN steps over the UNSW-NB15 and ISOT Botnet datasets, the prevention of fraudulent risks is accomplished. The ecological system as a whole is demonstrated in Figure 12.10.

The general structure of this cloud security framework based on ML is given in Figure 12.11. The striking feature of this framework is that, at different levels on the cloud network, it receives the benefits of intelligent decision construction and feedback mechanisms offered by machine learning classification models and artificial intelligence (AI). Based on the diagram, apart from protecting the information and the associated cloud framework using traditional antivirus software, firewalls of the operating system, etc., the suggested method provides artificial intelligence

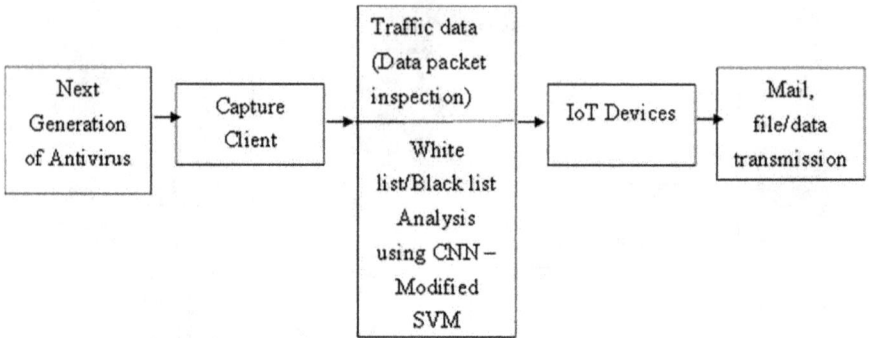

FIGURE 12.10 CNN-modified SVM cloud security system.

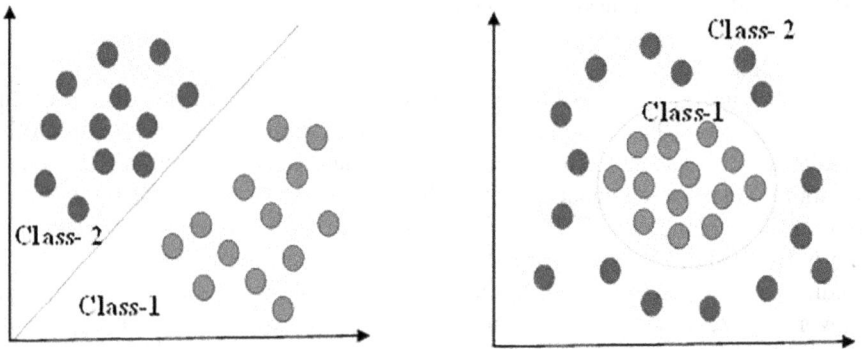

FIGURE 12.11 Support vector machine.

techniques of the next decade to collect and analyse customer feedback even at the point of entry level to establish that no unusual data reach the cloud-based framework. Protection must be recognized to be given at three stages, namely the point of entry where the customer demand is received, the network interface privacy, and the cloud framework and the end state where the user model of the system exists. Each one of the data streams is exposed to both encoding and decoding when the data are pre-processed and extracted into the device, and examined using ML algorithms to identify and warn unusual packets. These packets are then transmitted to the cloud layer, enabling the transmission of only the appropriate packets through the cloud network.

12.10.1 Modified Support Vector Machine

Different categories found in the datasets can be identified by the modified SVM classification model. The SVM grouping is shown in Figures 12.11 and 12.12, while Figure 12.11 displays the classification model of the modified SVM. The SVM classification model, based on the feature space, qualifies two or more groups. Each feature

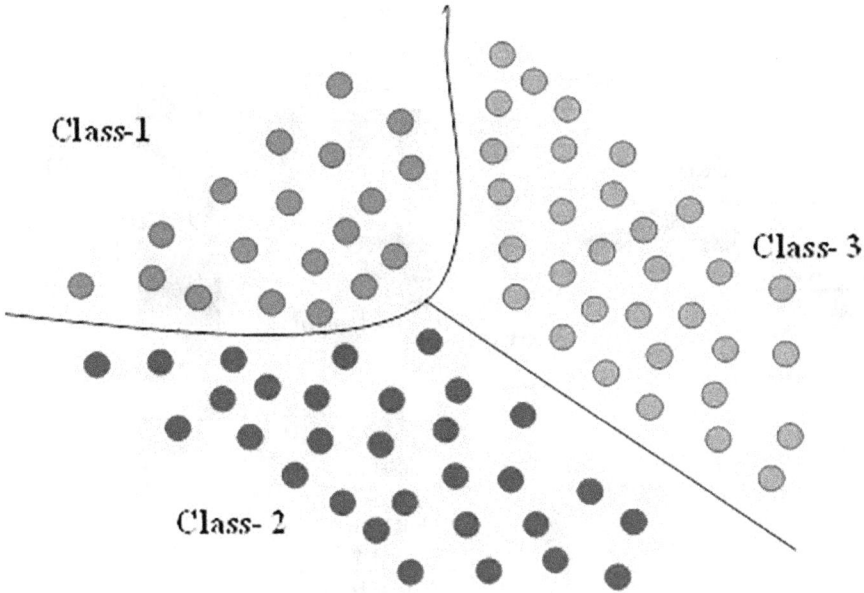

FIGURE 12.12   Modified support vector machine.

space in modified SVM qualifies the data into several groups as group 1, group 2, group 3, etc.

12.10.2   GPU-Based CNN Architecture

The GPU-based CNN framework involves the use of ML system-based categorization to identify any unusual activity in the cloud framework based on analysis of data. In this case, the GPU-based CNN framework is used for anomaly categorization. The cloud source data are fed to the CNN system. This involves introducing many levels of shared information as convolutional layer, and at each convolution point, the knowledge is converted into many types. The GPU-CNN output consists of various feature variables that act as the entry to the classification model of the modified SVM. The completely linked layer in the GPU-CNN typically identifies the final variable data where their length is two or three and eventually decides the unusual groups.

The inconsistency is divided into five types based on the cluster head, target base station, connectivity precision, timescale, and alert pattern. The performance vector quantity is also greater. Modified SVM is then added to the output variable coming from GPU-CNN to identify the abnormal groups. The confidential outcome received from GPU-CNN is well established to be highly precise. To classify the precise category of the data, the output variable received from the GPU-CNN is labelled using modified SVM. This modified SVM classifier is extensively qualified to classify activities across the cloud network depending on the relevant data, and the activities in the real-time cloud framework are classified as ordinary activities and abnormal

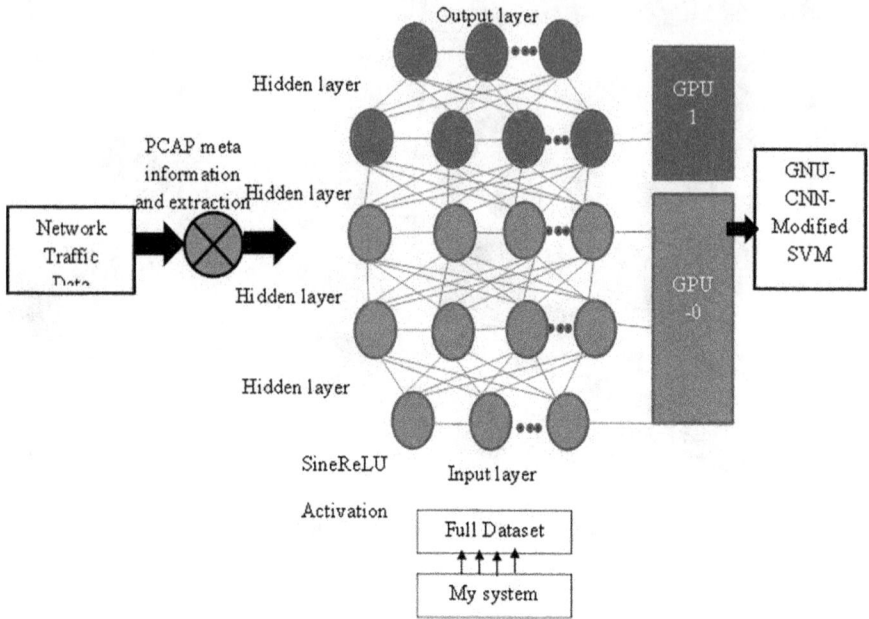

FIGURE 12.13 GPU-based CNN-modified SVM functionality.

activities when the extracted features are given as inputs. The GPU-CNN framework is demonstrated in Figure 12.13. The number of layers hidden is not restricted, but the number of layers hidden varies from 1 to 10. It is optimum when the amount of layer is 5. For the triggering of all the secret levels, SineReLU has been used. All the levels of matrices and consolidation are large in general, and the number of measurements is 250. When the data are interpreted in the GPU-CNN, the size of the data from the matrix to the variables is decreased. The actual variable is the chosen variable of the function added to the modified SVM to determine if the data are regular or irregular.

12.10.3 DATA TRAFFIC ANALYSIS FOR GPU-BASED CNN NETWORK

A test model of the dataset is given in Figure 12.14 for the complete data framework and the empirical set-up. The GPU-based convolution neural network has one hidden layer, one activation function, and a larger number of accumulating layer contortions. The entry field and the output field and the interface field are omitted as they affect the design. The collection of all raw data is fed into the hidden layers, where the information is transformed into arrays of convent. The convent and pooling levels also know the full characteristics of the results. All the convent accumulation levels are also called hidden layers, because they acquire and retrieve the entire hidden data from the source data after some approximation of operation.

The modified SVM feature activates the output unit. The modified SVM is used to make use of discrete identifiers such as 'normal' or 'abnormal' to identify performance results. In the GPU-CNN and modified SVM, the framework uses Keras and

SourceIP, Sourceport, DestinationIP, DestinationPort, Protocol, FlowBytes, Flowduration, FlowIATmean, FlowPackets, FlowIAT Max, FlowIATStd, FlowIATmin, FwdIATMean, FwdIATStd, FwdIATMax, FwdIATMin, BwdIATMean, BwdIATStd, BwdIATMax, BwdIATMin, ActiveStd, ActiveMean, ActiveMax, ActiveMin, IdleMean, IdleStd, IdleMax, IdleMin, ActiveStd, ActiveMean, ActiveMax, ActiveMin, IdleMean, IdleStd, IdleMax, IdleMin, Label, 10.0.2.15, 53912, 216.58.208.46, 80, 6, 434, 0, 4597.7011494252, 435, 0, 435, 435,0, 0, 0,0, 0, 0, 0, 0, 0, 0, NON-

FIGURE 12.14 Sample data.

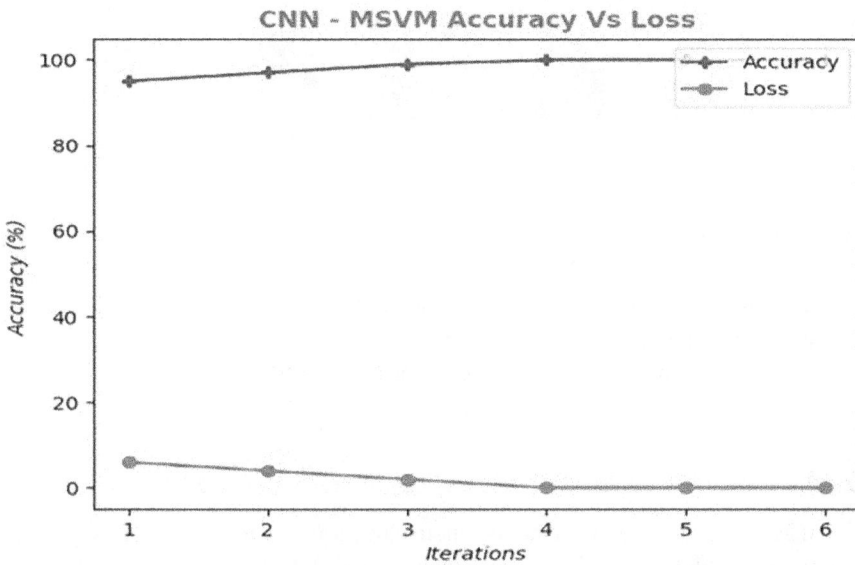

FIGURE 12.15 CNN-modified SVM accuracy vs loss.

TensorFlow to effectively train the CNN model. Finally, to improve the GPU-CNN, a cross-entropy is implemented to measure the loss. With multiple iterations, the suggested method is educated on many occasions. The specification of the GPU-based CNN-modified SVM model is shown in Figure 12.13 to increase the efficiency of determining anomalies and ordinary data with minimized failure for a larger amount of epochs. Compared with the different current supervised ML methods such as linear regression, support vector machine, naive Bayes classification, and random forest classification, the performance achieved using the GPU-based CNN-modified SVM approach is shown in Figure 12.15. Accuracy, recalling, and F-score, which assess the efficacy of the categorization skill, are the quality indicators compared to one another. The proposed GPU-based CNN-modified SVM is capable of detecting and defining the TOR-10 class. Similarly, GPU-based CNN is also necessary

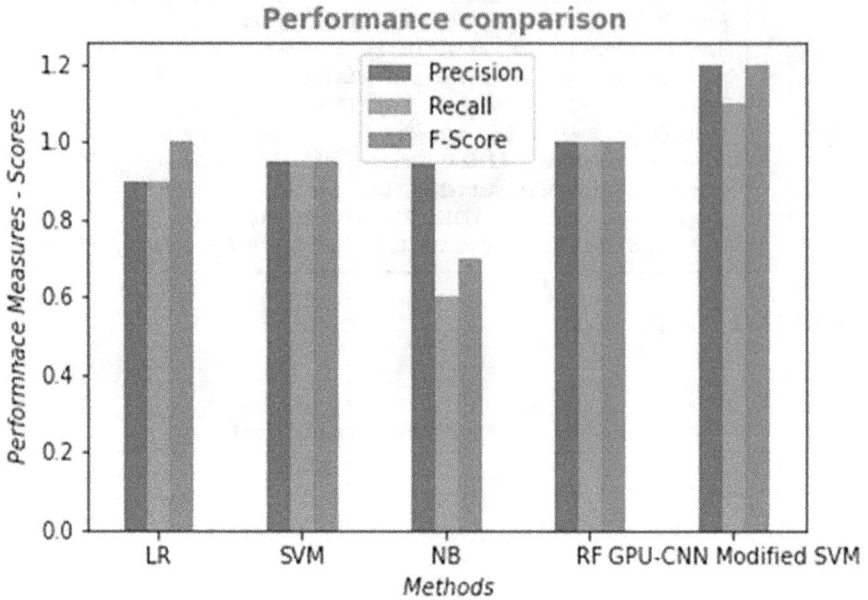

FIGURE 12.16 Performance comparison.

to be graded to identify the NON-TOR-10 class. It is reported from the confidential outcomes that the suggested GPU-based CNN-modified SVM approach decreases the amount of instances of function point for NON-TOR. The effect of contrast is illustrated in Figure 12.16.

12.10.4 BENEFITS OF AI AND MACHINE LEARNING FOR CLOUD SECURITY

AI and ML may not be a magic bullet, but in cloud security techniques, they can still perform a significant role. Cloud development indicates no indications of decreasing, according to Gartner, with 95% of businesses now utilizing it for at least a few of their activities. But many IT practitioners still identify the cloud as the key field of security vulnerabilities inside their organization, despite rising cloud acceptance, with 50% of businesses expecting to raise their cloud protection expenses over the next 12 months, a study from cyber security experts suggests. Some businesses are switching to AI and ML to boost their cloud protection to tackle this and reduce their risk of encountering a hack. Artificial intelligence is a technology that is identical to people and can solve tasks and learn on its own. A branch of AI that includes algorithm to improve from results is machine learning. The more statistical methods it evaluates, the more reliant on these anomalies it integrates and self-adjusts, and the more important its observations appear.

12.10.4.1 Big Data Processing

Cyber security systems create enormous volumes of data that any human team will ever peruse through and examine. Most of these data are used by ML technologies to identify attack issues. The more knowledge it processes, the more variations it identifies and observes, which it then applies to identify variations in the regular flow of pattern. Cyber risks may be these shifts. Machine learning, for instance, takes note of what is considered natural, such as when and where workers log into their systems, what they routinely access, and other patterns of traffic and user behaviours. Variations from these principles get reported, such as signing in during the early hours of the morning. This means that possible attacks can be emphasized and dispensed with in a quicker manner.

12.10.4.2 Event Prediction
By utilizing a more information-driven methodology, man-made brainpower can be utilized to identify and proactively alert on shortcomings and weakness both that are being misused at the present time and that may be abused later on. This works by examining information coming all through secured endpoints, both distinguishing dangers dependent on known conduct and spotting yet realized dangers dependent on prescient investigation. This more prescient methodology gathers all endpoint movement information as opposed to simply the 'terrible' action and improves it from different sources to help address the main derivers of an expected assault, instead of simply limiting the impacts once an assault is distinguished. It can likewise help make a more limited cycle among recognition and remediation by guaranteeing a security group can respond quicker with better information.

12.10.4.3 Event Detection and Blocking
At the point when AI and AI advancements measure the information produced by the frameworks and discover irregularities, they can either alarm a human or react by closing a particular client out, among different alternatives. By making these strides, occasions are regularly identified and hindered inside hours, closing down the progression of conceivably perilous code into the organization and forestalling an information spill. This cycle of inspecting and relating information across geology progressively empowers organizations to possibly get long periods of caution and time to make a move in front of security occasions.

12.10.4.4 Delegating to Automated Technologies
Alarms about possible dangers or abnormalities are extremely normal with numerous security stages, yet there is a great deal of potential with mechanized advances to wipe out a ton of the clamour to have the option to zero in on the significant things. At the point when security groups have AI and AI innovations dealing with routine errands and first-level security examination, they are allowed to zero in on more basic or complex dangers. This is particularly significant given that the current abilities lack in network protection. With 51% of associations professing to have a tricky lack of network protection abilities, organizations can calm a portion of the weight by designating the primary degree of examination to bots, permitting security experts to zero in their endeavours on combating more troublesome assaults. This doesn't mean these innovations can supplant human experts, as digital assaults frequently start from both human and machine endeavours and hence require reactions from both the

people and machines, too. Notwithstanding, it permits examiners to organize their undertaking all the more effectively.

12.11 CONCLUSIONS

Significant developments in information technologies have resulted in new cyber security problems arising. New methods that are more efficient, customizable, and versatile involve the mathematical sophistication of cyberattacks. This chapter focuses on applying the AI-based approach to cyber security problems. Specifically, in detecting attacks, vulnerability scanning, and other realms, such as spam filtering and malware identification, the AI is applied. Recently, IDS, anomaly detection and categorization, spoofing, and email attachments have been the key targets for AI implementation in cyber security. The implementation of DL has increasingly become the key phenomenon in those regions. In addition, along with ML/DL, the mixture of other sophisticated approaches, such as bio-inspired approaches, also stimulated investigators' interest. Such variations yield findings that are very positive and establish a tendency for more study. While AI's role in overcoming cyber security issues remains to be studied, some of the issues concerning the implementation of AI-based defences are also striking. The hostile assault towards AI techniques, for example, is the advent of intelligent software malware. The study on the discovery of responses to these challenges should also be further examined. Hence, enhanced ML method such as GPU-based CNN-modified SVM is used for ensuring cloud protection. The test is performed by introducing all the input data into the GPU-based CNN model's hidden layers. The key focus of this work is to establish a better learning method to cloud protection for cloud computing. It also focuses on the implementation of comprehensive scientific work using GPU-based CNN that is combined with the modified SVM approach. It is established and inferred from the experimental results and evaluation of output that the expanded supervised ML methods are particularly appropriate and relevant in real-time cloud services. The performance is tested by experiments on different datasets, and it is confirmed that the GPU-based CNN-modified SVM method performs better.

REFERENCES

1. Berman, D. S., Buczak, A. L., Chavis, J. S., and Corbett, C. L. A survey of deep learning methods for cyber security. *Information* 2019, 10, 122.
2. Paul, S., Jain, R., Samaka, M., and Pan, J. Application delivery in multi- cloud environments using software defined networking. *Computer Networks Special Issue on Cloud Networking and Communications*, 2014, 68, 166–186.
3. Karn, R. R., Kudva, P., and Elfadel, I. A. Dynamic autoselection and autotuning of machine learning models for cloud network analytics. *IEEE Transactions on Parallel Distributed Systems* 2019, 30(5), 1052–1064.
4. Buczak, A. L. and Guven, E. A survey of data mining and machine learning methods for cyber security intrusion detection. *IIEEE Communications Surveys & Tutorials* 2015, 18, 1153–1176.
5. Kaufman, L. M. Data security in the world of cloud computing. *IEEE Security & Privacy* 2009, 7(4), 61–64.

6. Borisenko, K., Rukavitsyn, A., Gurtov, A., and Shorov, A. Detecting the origin of DDoS attacks in OpenStack cloud platform using data mining techniques, Internet Things, Smart Spaces, Next Gener. *Network System* 2016, 9870, 303–315.

7. Khandelwal, P. and Sudhir, K. Introduction to Artificial Intelligence and its Applications. On Emerging Trends in Information Technology (NCETIT'2018) with the Theme-'The Changing Landscape of Cyber Security: Challenges, 2018, 94.

8. Fernandes, D., Soares, L., Gomes, J., Freire, M., and Inácio, P. Security issues in cloud environments: A survey. *International Journal of Information Security* 2014, 13(2), 113–170.

9. Kamtam, A., Kamar, A., and Patkar, U. C. Artificial intelligence approaches in cyber security. *International Journal on Recent and Innovation Trends in Computing and Communication* 2016, 4(4), 5–9.

10. Torres, J. M., Comesaña, C. I., and García-Nieto, P. J. Machine learning techniques applied to cybersecurity. *International Journal of Machine Learning and Cybernetics* 2019, 10, 1–14.

11. Pandey, M. Artificial intelligence in cyber security. On Emerging Trends in Information Technology (NCETIT'2018) with the Theme-'The Changing Landscape of Cyber Security: Challenges, 2018, 66.

12. Bhamare, D., Salman, T., Samaka, M., Erbad, A., and Jain, R. Feasibility of supervised machine learning for cloud security. *In 2016 International Conference on Information Science and Security (ICISS)*, Pattaya, 2016, 1–5.

13. Apruzzese, G., Colajanni, M., Ferretti, L., Guido, A., and Marchetti, M. On the effectiveness of machine and deep learning for cybersecurity. *In 2018 10th International Conference on Cyber Conflict (CyCon)*, IEEE, 2018, 371–390.

14. Sommer, R. and Paxson, V. Outside the closed world: On using machine learning for network intrusion detection. *In IEEE Symposium on Security and Privacy*, 2010, 305–316.

15. Vidhya, V. A review of DOS attacks in cloud computing. *IOSR Journal of Computer Engineering (IOSRJCE)*, 2014, 16(5), 32–35. https://ieeexplore.ieee.org/document/5504793

16. Ye, Y., Chen, L., Hou, S., Hardy, W., Li, X., and Deep, A. M. A heterogeneous deep learning framework for intelligent malware detection. *Knowledge and Information Systems* 2018, 54, 265–285.

17. Modi, C. et al. A survey of intrusion detection techniques in cloud. *Journal of Network and Computer Applications* 2013, 36(1), 42–57.

18. Nour, M. and Slay, J. The evaluation of network anomaly detection systems: Statistical analysis of the UNSW-NB15 data set and the comparison with the KDD99 data set. *Information Security Journal: A Global Perspective* 2016, 25, 1–14.

19. Aiwan, F. and Zhaofeng, Y. Image spam filtering using convolutional neural networks. *Personal and Ubiquitous Computing* 2018, 22(5–6), 1029–1037.

20. Li, J., Sun, L., Yan, Q., Li, Z., Srisa-an, W., and Ye, H. Significant permission identification for machine-learning-based android malware detection. *IEEE Transactions on Industrial Informatics* 2018, 14, 3216–3225.

21. Katoua, H. S. Exploiting the data mining methodology for cyber security. *Egyptian Computer Science Journal*, 2013, 37(6), 1–9.

22. Peddabachigari, S., Abraham, A., and Thomas, J. Intrusion detection systems using decision trees and support vector machines. *International Journal of Applied Science and Computations*, 2004, 11(3), 118–134.

23. Karbab, E.B., Debbabi, M., Derhab, A., and Mouheb, D. MalDozer: Automatic framework for android malware detection using deep learning. *Digital Investigation* 2018, 24, S48–S59.

24. Michalski, R., Carbonell, J., and Mitchell, T. *Machine Learning: An Artificial Intelligence Approach*. Springer Science & Business Media, Berlin, Germany, 2013.
25. Kumar, T. D., Samuel, T. A., and Kumar, T. A. Transforming 2 green cities with IoT. In: Saravanan, K. and Sakthinathan, G. (Eds) *Handbook of Green Engineering Technologies for Sustainable Smart Cities*, 17. CRC Press, Boca Raton, FL, 2021.

13 Enhanced Hybrid and Highly Secure Cryptosystem for Mitigating Security Issues in Cloud Environments

Hamid Ali Abed AL-Asadi
Iraq University College
University of Basrah

Amer S. Elameer
University of Information Technology and Communications

CONTENTS

13.1 INTRODUCTION

Rapid developments in information and communication domain have recently attracted many researchers to contribute more to this particular domain. Research works mainly concentrate on security issues, secure data storage, and various algorithms that focus mainly on attaining data privacy and data protection [1]. A complete security mechanism for attaining secure data exchange between environments is still a serious concern. Cloud computing is one of the leading data storage and data exchange mechanisms that is currently under consideration by numerous ongoing researchers. Almost all the organizations make use of cloud computing either directly or indirectly; therefore, this domain is mainly concentrated by researchers to propose numerous secure algorithms toward cloud computing. Even though cloud computing delivers finest techniques for improving information technology (IT)

DOI: 10.1201/9781003219880-13

possessions with reduced cost, better flexibility and higher productivity, this domain is still under concern which was not obviously defined by the researchers [2]. Various models were enacted by cloud computing to deliver software as a service [3], platform as a service [4] and infrastructure as a service [5], where the customers pay for usage in contrast to own resources. The main feature here in cloud computing is that it is well suited toward shared delivery of services. Recommendations from NIST introduced a service architecture that specifies them as collective service, application, data and infrastructure that was formed as a collection of computers, networks, data and storage reserves.

The components involved in cloud computing shall be easily provisioned and quickly implemented with better scalability features [5]. Therefore, cloud computing is observed to be greatly secure, private, offering data integrity, guarding property rights and rectifying other issues that are concerned with data storage in cloud. Numerous research works have been performed and implemented over cloud computing domain so as to attain data security and privacy in cloud environment. For example, one major solution that was identified in real time was the assignment of Dropbox to each and every user; therefore, Dropbox is observed as one of the leading data saving services in cloud. Here, the users can store any amount of data into the cloud and they can access their data anywhere/anytime using the Internet.

Another solution formulated toward data sharing in cloud environment is the usage of virtual machine (VM) that operates its specific operating systems behaving like physical machines [6–9]. Subsequently, the next method is the employment of encryption techniques, where the sender sends the encrypted data to the cloud server for storage; by this way, the security and privacy of the stored information can be well maintained. Some of the security and data privacy preserving encryption algorithms currently available are Amazon EC2, AES-256, Cipher Cloud, Cupertino, etc. Even though numerous algorithms were formulated by the researchers to offer better privacy and security, few parameters have yet to be considered that were not discussed in their research. On the basis of the requirements and necessities of the organizations or users, there are four main deployment models possessing their own features that are to be considered and could be understood from Figure 13.1.

The first one is the private cloud, which is also called internal cloud. These clouds are formulated and maintained by one specific party or one specific company/ organization, thereby maintaining privacy through application programming interface (API). The second one is the public cloud, which is also referred to as external cloud. This cloud is formulated by external business concerns, and these clouds are left out to make it available to the public. The third one is the community clouds, which are formulated and maintained by some group of companies that have similar service requirements; in other words, these clouds are established and maintained by companies that support specific communities. The fourth category is the hybrid environment, which is the grouping of two or more of the aforementioned types. The main reason to form this type of cloud is to attain resource sharing among various deployment models; hence, data or applications can be transferred from one type of cloud to the other. The cloud data management in this type of hybrid clouds is done by both internal parties and some third parties externally. Since the hybrid cloud possesses more essential features when compared to the other three clouds, this chapter

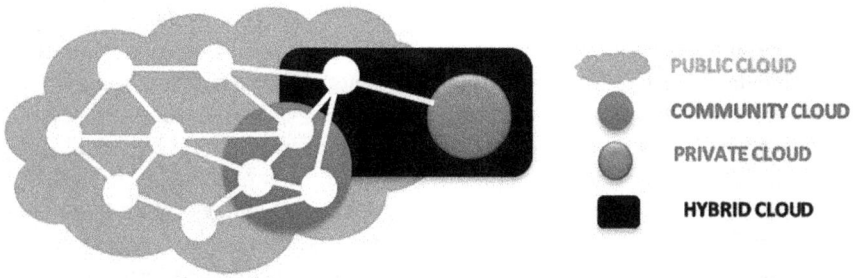

PUBLIC CLOUD

COMMUNITY CLOUD

PRIVATE CLOUD

HYBRID CLOUD

FIGURE 13.1 Deployment models in cloud environment [10].

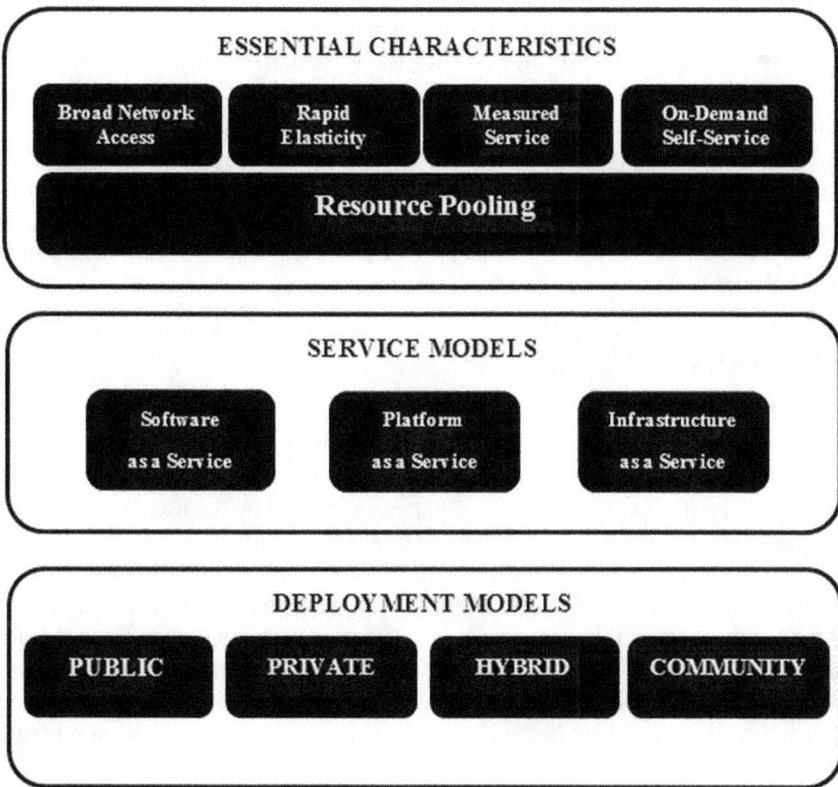

ESSENTIAL CHARACTERISTICS

Broad Network Access

Rapid Elasticity

Measured Service

On-Demand Self-Service

Resource Pooling

SERVICE MODELS

Software as a Service

Platform as a Service

Infrastructure as a Service

DEPLOYMENT MODELS

PUBLIC   PRIVATE   HYBRID   COMMUNITY

FIGURE 13.2 NIST model and basic layers in cloud computing.

will be concentrating on the formulation of a data security mechanism for hybrid clouds.

Hybrid clouds possess variable combinations of public, private and community clouds. The essential characteristics, services and information provided can be summarized by the employment of NIST service architecture. Based on NIST, Figure 13.2

shows the basic layers that are involved in cloud computing methods. Initially, the service provider of cloud will have to decide the type of deployment model that needs to be selected based on the nature of requirements of the customers. Apart from the deployment model selected, each and every cloud service will possess five essential characteristics that are mentioned in the upper layer as shown in Figure 13.2. This NIST model specifies cloud computing as mechanisms or models for on-demand accessing of resources from a common pool of shared resources with minimal intervention by the cloud service providers. As per these above NIST specifications, a model called NIST Cloud Computing Reference Architecture was formulated by the working groups, which illustrates a universal higher-level theoretical model to discuss the necessities, structures and procedures involved in cloud computing.

This possesses a group of interpretations and explanations that form the basement for entailing the features, usage and standard toward cloud computing, thereby correlating with the companion cloud computing taxonomy.

The remaining sections in this chapter are systematized as follows: A detailed literature review of the existing approaches toward enhancing security in cloud computing by entailing their features, advantages and disadvantages is summarized in Section 13.2. The framework and functionalities of the proposed security model for cloud environments are discussed in Section 13.3. Section 13.4 highlights the simulation results and provides a discussion, thereby elaborating the improvements in the proposed algorithm, which is contrasted with its peers. Finally, Section 13.5 gives the conclusions.

13.2   REVIEW OF LITERATURE

Researchers have proposed attribute-based encryption (ABE), which is a demonstrable information security technique for providing information privacy in cloud systems [11–13]. In their approach, information decryption algorithm is formulated by considering the nature of attributes that are requested by the users in association with the encrypted information. The main drawback faced by their approach is that the cloud service providers face greater amount of computation overheads and storage overheads while verifying user attribute with outsourced encrypted information. By the employment of third-party auditors, there is possibility to trim down the storage overhead, computation overhead and communication overhead at the cloud servers, thereby offering better efficiency in storing the information at the cloud server. Research works were done by authors to formulate an approach that incorporates six dissimilar symmetric key RSA algorithms so as to show betterment toward data storage in cloud environments [14]. Their method incorporated two servers: one for data thereby called data server and the other for key thereby called key server, and at the client side, as usual, encryption and decryption of the data take place. Creation of increased overheads at the cloud server due to the presence of two servers is the main drawback faced by this approach. Efforts were made by researchers to propose a methodology that describes cloud as a distributed design that integrates server resource over scalable platforms, thereby providing on-demand computing resources and services [15]. Therefore, they have described cloud to be a variable platform in companies for building their infrastructure in a decent manner. When companies take advantage of cloud toward information storage at the cloud servers, they will have to face serious tasks toward reconsidering the present security approach. The authors in Ref. [16] proposed a structure for cloud environments that are capable of delivering on-demand provisions toward data storage. The three main layers considered by the authors in their approach are the input, the policy and the security layers. In the first layer, three distinct checks were performed so as to boost the security features. In the second layer, based on the needed security provisions, concerned security parameters and privacy parameters are implemented. In the third layer, different security mechanisms are incorporated at the different domains.

Research works carried out in Ref. [17] brought out a mechanism that controls the data access from the cloud servers. In their approach, the authors employed a distributed access control strategy that employs multiple key distribution centers (KDCs) for sharing the keys to the users. In most of the prevailing strategies, centralized access control mechanisms were employed, which in fact leads to single point failures. In this approach, they considered that a user who is blocked by the server once will not be able to steal any information from the cloud server. Also, the identity of that particular user will not be disclosed by the cloud server. In this approach, SHA-based encryption algorithm is employed for hiding the user data. Here, the cloud server clearly knows the access mechanism that is followed by each and every user in that particular cloud environment.

An enhanced research work was done by authors in Ref. [18] for formulating a cloud strategy that improves data search and user search capabilities by the employment of a secure fuzzy-based ranking keywords searching scheme. Here, in this approach, the results clearly demonstrated that the overhead in generating the trapdoor for each query has greatly been trimmed down. The main drawback faced by this approach is the increase in search time when compared to other methods despite their targeted advantages. Hence, there is a greater demand in formulating a secure mechanism that increases the data privacy, data security and reduced searching time duration.

The authors in Ref. [19] proposed an authentication-based key exchange strategy for cloud computing environment, which was typically referred to as cloud computing background key exchange (CCBKE). It mainly targeted secure scheduling of real-world application scenarios when dealing with hybrid cloud computing environments. This system was mainly implemented on the basis of frequently employed Internet Key Exchange (IKE) strategy and randomness reuse approach. Therefore, from these literature studies, we could see that there is a wider scope in formulating a secure algorithm that decreases data encryption and decryption times and increases data security while storing and accessing the data from the cloud servers, which are being discussed further in subsequent sections.

13.3   PROPOSED ENHANCED HPS ALGORITHM

To mitigate the issues related to security in cloud computing, we have proposed a hybrid secure algorithm that is referred to as enhanced hybrid privacy and secure (EHPS) algorithm. The proposed EHPS algorithm is based on the hybrid approach that incorporates the features of AES, DES, CBC and Triple DES methods. The process of encrypting and decrypting the data is performed by the employment of AES and hybrid approach. Key management is accomplished by the employment of the concept of group key management algorithm. For the purpose of key management and key exchange, key is generated by the use of randomly encrypted key generation algorithm and key agreement is accomplished with the help of Diffie–Hellman. Initially, we employ the combination of AES and hybrid approach for attaining symmetric encryption. The combination of these two will result in the formation of the proposed EHPS algorithm that reduces the encryption time when compared to the original AES algorithm. The resultant total time of the proposed algorithm will be the summation of the run-times of AES and the hybrid approach. The hybrid algorithm which is the combination of DES, CBC and Triple DES approaches requires lesser run-time when compared to the existing algorithm. In addition to the reduced encryption and decryption times, this hybrid algorithm attains a better security mechanism. The overall system architecture at the sender side is elaborated in Figure 13.3. At the sender side, the plaintext is encrypted with AES and hybrid algorithm with the use of vector key that is used by the sender during encryption. This encryption key is encrypted once again and is finally transported to the channel via key management approach.

At the sender side, the following steps are followed for attaining increased security:

The overall system architecture at the receiver end with functional representation is shown in Figure 13.4. The receiver side employs two decryption algorithms AES and hybrid decryption algorithm for decrypting the encrypted data received at the receiver module. The key encrypted using the randomly encrypted key generation algorithm is decrypted by the employment of decryption part of the same algorithm.

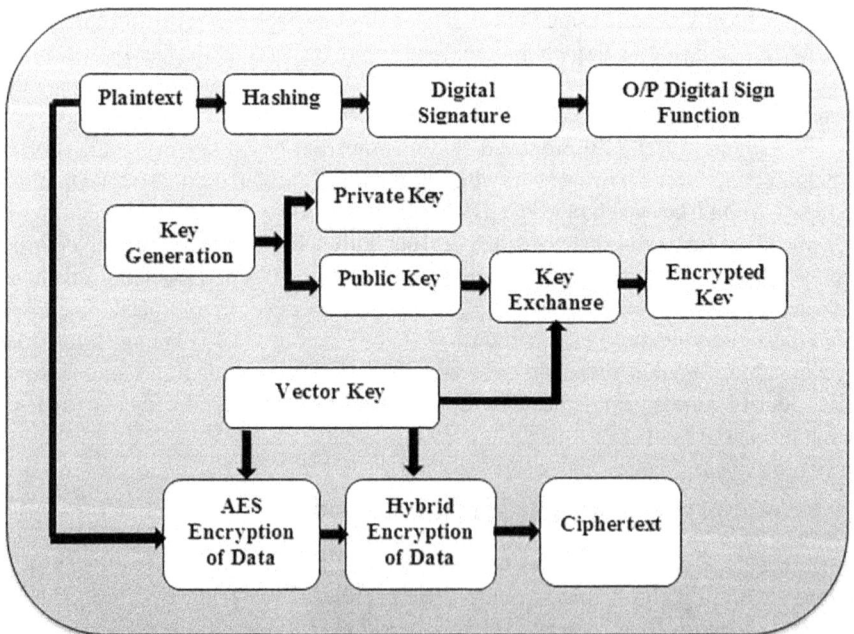

FIGURE 13.3   System architecture at sender side.

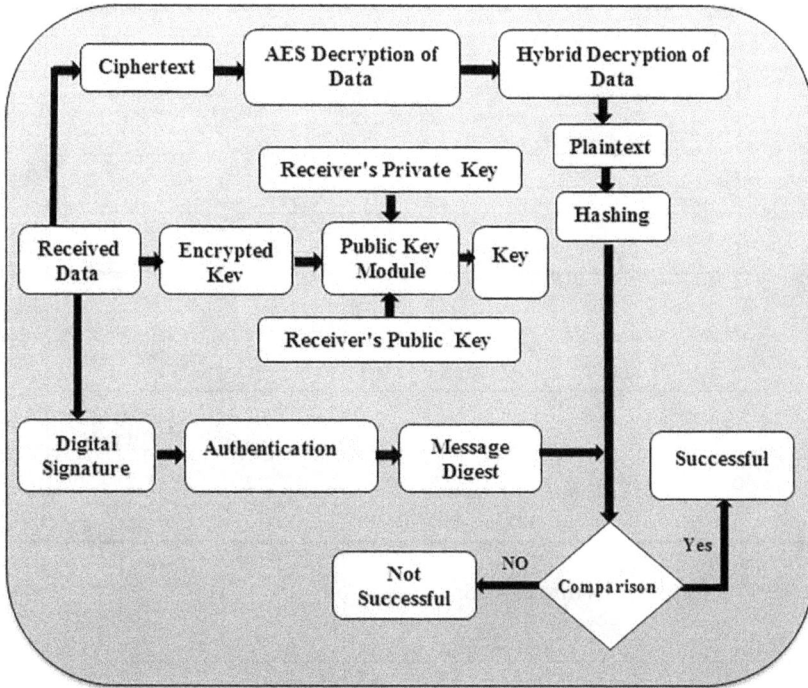

FIGURE 13.4 System architecture at receiver side.

The key corresponds to the private key of the sender and the public key of the receiver, and on the application of these keys, the plaintext can be extracted. Subsequently, the message digest is applied over the plaintext. This generated message digest will be compared with 128-bit message digest, and if both these message digests are identical, then the data will be accepted or else the information is discarded. The above steps are followed at the receiver side.

13.4 RESULTS AND DISCUSSION

The security effect of the cryptosystem can be increased by the combination of AES and hybrid security mechanisms that result in the proposed EHPS algorithm. The proposed strategy mainly concentrates on security and is a solution for data security issues in storage systems. The proposed EHPS algorithm is based on the hybrid approach that incorporates the features of AES, DES, CBC and Triple DES algorithms. The plaintext file is split into three to six portions as per the requirements of the user input. Every part of the file is encrypted by using the aforementioned algorithms.

13.4.1 TEST FILES

The testing of the proposed approach was done over different file sizes varying from 1 to 30 MB. The encryption time of the proposed EHPS algorithm was carefully

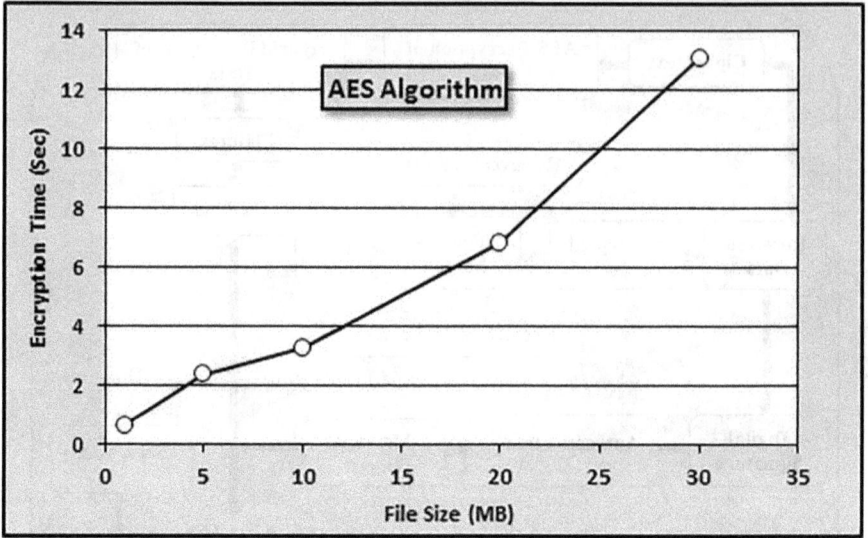

FIGURE 13.5   Encryption time of AES.

TABLE 13.1
AES Encryption Time

File Size (MB)	Encryption Time (seconds)—AES Algorithm
1	0.617
5	2.344
10	3.257
20	6.829
30	13.110

noted down and compared with the AES algorithm. Also, the decryption time of the proposed EHPS algorithm was clearly investigated and compared with the existing AES algorithm. Moreover, the encryption and decryption of the proposed EHPS algorithm were compared with hybrid cryptosystem.

Figure 13.5 and Table 13.1 show the analytical results of encryption times of AES when varying file sizes of 1–30 MB are taken for analysis. From the results, it is seen that the encryption time exponentially increases when file sizes are increased. This scenario is common in almost every cryptographic algorithm, as the file size is directly proportional to the encryption time, but an algorithm that reduces this exponential growth to a considerable limit is greatly preferred.

Figure 13.6 and Table 13.2 show the simulation results in terms of decryption time with respect to varying file sizes of 1, 5, 10, 20 and 30 MB. The same situation prevails here. One can see the exponential growth in decryption time when the file sizes are increased from lower end to the upper end. Even though this is a common

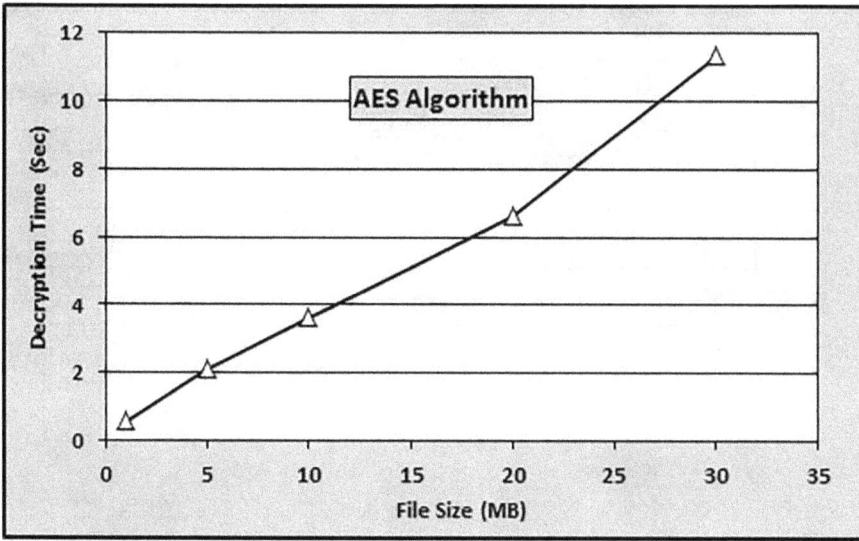

FIGURE 13.6 Decryption time of AES.

TABLE 13.2
AES Decryption Time

File Size (MB)	Decryption Time (seconds)—AES Algorithm
1	0.589
5	2.099
10	3.643
20	6.625
30	11.345

scenario, a methodology or algorithm that trims down the exponential growth with respect to the increase in data size will be well favored.

Figure 13.7 and Table 13.3 show the comparative analysis between the encryption results attained using the AES algorithm and the proposed EHPS algorithm. The encryption times of AES when 1, 5, 10, 20 and 30 MB data were used are 0.617, 2.344, 3.257, 6.829 and 13.110 seconds, respectively.

Consequently, the encryption times of the proposed approach for the same data files are 0.506, 1.721, 2.832, 6.173 and 11.990 seconds, respectively. The average encryption time of AES is 5.2314 seconds, while the average encryption time of the proposed algorithm is 4.6444 seconds.

As a result, the proposed EHPS algorithm shows 11.22% improvement when compared to AES. In the proposed approach, data encryption and decryption are performed by the utilization of AES and hybrid approach. Also, key management is attained by the concept of group key management algorithm.

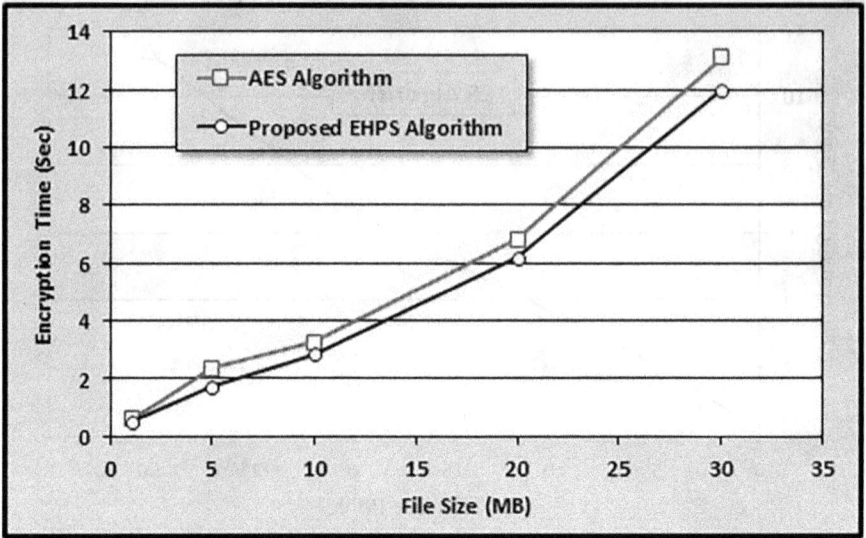

FIGURE 13.7 Scenario of hybrid encryption time.

TABLE 13.3
Hybrid Encryption Time

File Size (MB)	Encryption Time (seconds)	
	AES Algorithm	**Proposed EHPS Algorithm**
1	0.617	0.506
5	2.344	1.721
10	3.257	2.832
20	6.829	6.173
30	13.110	11.990

Furthermore, for the purpose of key management and key exchange, key is generated by the use of randomly encrypted key generation algorithm. These are the reasons for the proposed algorithm to exhibit a decent reduction in encryption time.

Figure 13.8 and Table 13.4 show the scenario of variation in decryption time of this suggested approach when compared to AES algorithm. These values of decryption times of both approaches were recorded and tabulated for file sizes of 1, 5, 10, 20 and 30 MB. For these above values of file sizes, the decryption times of AES were recorded as 0.589, 2.099, 3.643, 6.625 and 11.345 seconds, respectively. Therefore, the AES exhibits an average decryption time of 4.8602 seconds. Similarly, for the above values, the decryption times of the proposed algorithm were recorded as 0.504, 1.853, 3.172, 6.004 and 10.886 seconds, respectively. Hence, the proposed algorithm exhibits an average decryption time of 4.4838. As a result, the proposed approach reveals an 8.39% reduction in decryption time when compared to the AES algorithm.

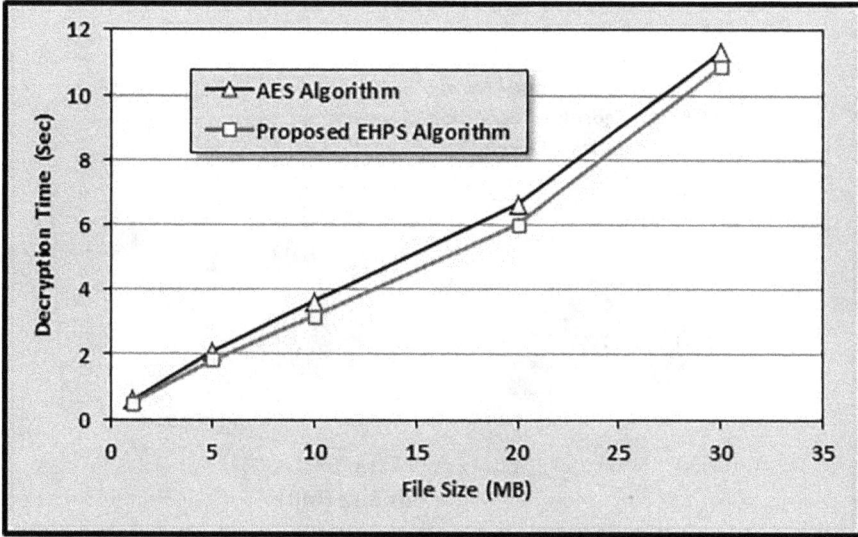

FIGURE 13.8 Scenario of hybrid decryption time.

TABLE 13.4
Hybrid Decryption Time

File Size (MB)	Decryption Time (seconds)	
	AES Algorithm	Proposed EHPS Algorithm
1	0.589	0.504
5	2.099	1.853
10	3.643	3.172
20	6.625	6.004
30	11.345	10.886

Figure 13.9 and Table 13.5 show the scenario of encryption and decryption algorithms. The proposed EHPS algorithm incorporates the features of AES, DES, CBC and Triple DES approaches. The proposed hybrid approach requires 11.22% reduced encryption time when compared to the AES algorithm encryption. Similarly, the proposed hybrid approach exhibits 8.39% reduced decryption time when compared to AES algorithm decryption. It could be inferred that, with any single algorithm, such kind of increased data security cannot be delivered; hence, when data security is a major concern, it is apt to move toward hybrid approaches.

Comparative analysis of encryption times for files sizes varying from 1 to 30 MB of different algorithms that include EHPS algorithm and hybrid cryptosystem is shown in Figure 13.10 and Table 13.6. The average encryption time of EHPS algorithm is 4.6444 seconds, and the average encryption time of hybrid cryptosystem is 5.6602 seconds.

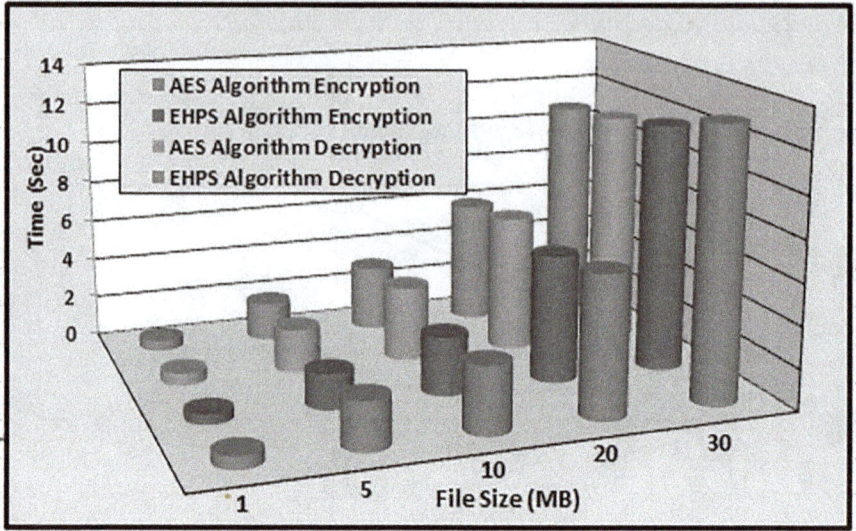

FIGURE 13.9   Scenario of encryption and decryption algorithms.

TABLE 13.5

Comparison of Encryption and Decryption Algorithms

File Size (MB)	Encryption Time (seconds)		Decryption Time (seconds)	
	AES Algorithm	Proposed EHPS Algorithm	AES Algorithm	Proposed EHPS Algorithm
1	0.617	0.506	0.589	0.504
5	2.344	1.721	2.099	1.853
10	3.257	2.832	3.643	3.172
20	6.829	6.173	6.625	6.004
30	13.110	11.990	11.345	10.886

The proposed algorithm displays 17.94% reduced encryption time when compared to hybrid cryptosystem. Hence, in cases when the encryption time is a major concern, the proposed EHPS algorithm could be preferred when compared to hybrid cryptosystem.

Comparative analysis of decryption times for files sizes varying from 1 to 30 MB of different algorithms that include EHPS algorithm and hybrid cryptosystem is shown in Figure 13.11 and Table 13.7. The average decryption time of the proposed algorithm was recorded as 4.4838. Similarly, the average decryption time of hybrid cryptosystem was recorded as 5.2202 seconds. The proposed approach exhibits 14.10% reduced decryption time when compared to hybrid cryptosystem. Hence, in cases when the decryption time is a serious concern, the proposed EHPS algorithm can be opted for implementation.

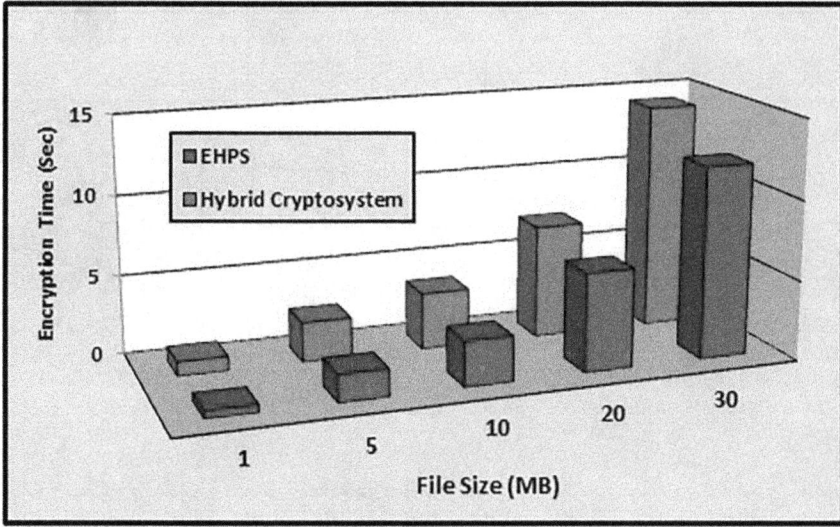

FIGURE 13.10   Graphical representation of EHPS encryption and hybrid cryptosystem.

TABLE 13.6
Comparison of EHPS Encryption and Hybrid Cryptosystem

File Size (MB)	EHPS Encryption	Hybrid Cryptosystem
1	0.506	0.889
5	1.721	2.499
10	2.832	3.543
20	6.173	7.125
30	11.990	14.245

13.5   CONCLUSIONS

The major reason behind the application of cloud computing in numerous real-world environments is its increased scalability, increased flexibility, enhanced agility and simplicity. Various research works have been executed and instigated over cloud computing domain so as to accomplish information integrity over cloud environment.

1. In this chapter, a hybrid algorithm for enhancing the security over cloud environments, referred to as EHPS algorithm, has been proposed and well evaluated among its peers. The proposed algorithm incorporates the features of AES, DES, CBC and Triple DES algorithms.
2. The incorporation of higher-end security features of these approaches makes the proposed hybrid algorithm show reduced encryption/decryption times. The proposed algorithm shows 11.22% reduced encryption time and 8.39% reduced decryption time when compared to AES.

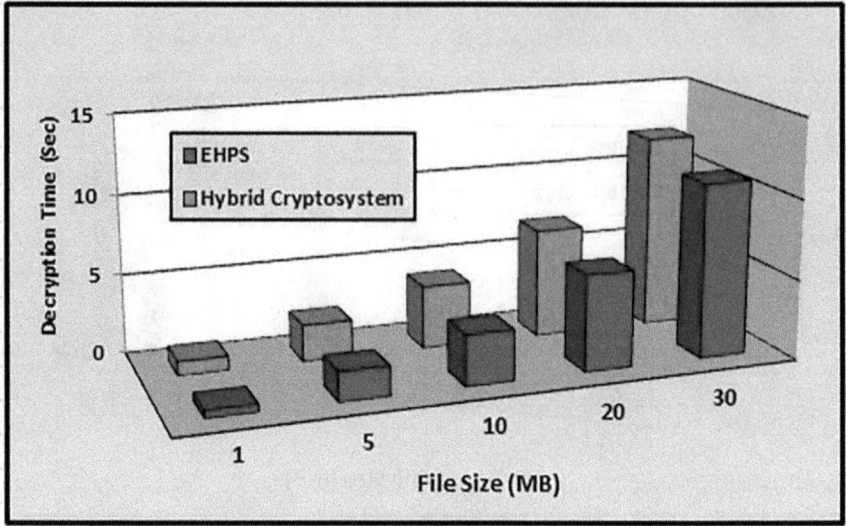

FIGURE 13.11   Graphical representation of EHPS decryption and hybrid cryptosystem.

TABLE 13.7
Comparison of EHPS Decryption and Hybrid Cryptosystem

File Size (MB)	EHPS Decryption	Hybrid Cryptosystem
1	0.504	0.889
5	1.853	2.299
10	3.172	3.943
20	6.004	6.825
30	10.886	12.145

3. Also, the proposed approach shows reduced encryption/decryption times when compared to hybrid cryptosystem that will make the proposed algorithm as a viable choice for many real-world applications in future.

REFERENCES

1. Sun, P. J. (2020). Security and privacy protection in cloud computing: Discussions and challenges. *Journal of Network and Computer Applications*, 102642. doi: 10.1016/j.jnca.2020.102642.
2. Sunyaev, A. (2020). Cloud computing. In: *Internet Computing*. Springer, Cham. doi: 10.1007/978-3-030-34957-8_7.
3. Rumale, A. S., & Chaudhari, D. N. (2017). Cloud computing: Software as a service. *2017 Second International Conference on Electrical, Computer and Communication Technologies (ICECCT)*. doi: 10.1109/icecct.2017.8117817.

4. Dhuldhule, P. A., Lakshmi, J., & Nandy, S. K. (2015). High performance computing cloud: A platform-as-a-service perspective. *2015 International Conference on Cloud Computing and Big Data (CCBD)*. doi: 10.1109/ccbd.2015.56.

5. Yin, X., Chen, X., Chen, L., & Li, H. (2020). Extension of research on security as a service for VMs in IaaS platform. *Security and Communication Networks*, 2020, 1–16. doi: 10.1155/2020/8538519.

6. Jalali Moghaddam, M., Esmaeilzadeh, A., Ghavipour, M., & Zadeh, A. K. (2020). Minimizing virtual machine migration probability in cloud computing environments. *Cluster Computing*. doi: 10.1007/s10586-020-03067-5.

7. Pang, S., Xu, K., Wang, S., Wang, M., & Wang, S. (2020). Energy-saving virtual machine placement method for user experience in cloud environment. *Mathematical Problems in Engineering*, 2020, 1–9. doi: 10.1155/2020/4784191.

8. Azizi, S., Zandsalimi, M., & Li, D. (2020). An energy-efficient algorithm for virtual machine placement optimization in cloud data centers. *Cluster Computing*. doi: 10.1007/s10586-020-03096-0.

9. Matilda, S., & Kumar, T. A. (2020). The winning combo: Cryptocurrency and blockchain. In *Blockchain Technology*, pp. 199–217. Boca Raton, FL: CRC Press.

10. Savu, L. (2011). Cloud computing: Deployment models, delivery models, risks and research challenges. *2011 International Conference on Computer and Management (CAMAN)*. doi: 10.1109/caman.2011.5778816.

11. Horváth, M. (2015). Attribute-based encryption optimized for cloud computing. *SOFSEM 2015: Theory and Practice of Computer Science*, 566–577. doi: 10.1007/978-3-662-46078-8_47.

12. Sumathi, M., & Sangeetha, S. (2020). A group-key-based sensitive attribute protection in cloud storage using modified random Fibonacci cryptography. *Complex & Intelligent Systems*. doi: 10.1007/s40747-020-00162-3.

13. Song, Y., Wang, H., Wei, X., & Wu, L. (2019). Efficient attribute-based encryption with privacy-preserving key generation and its application in industrial cloud. *Security and Communication Networks*, 2019, 1–9. doi: 10.1155/2019/3249726.

14. Rohini, & Sharma, T. (2018). Proposed hybrid RSA algorithm for cloud computing. *2018 2nd International Conference on Inventive Systems and Control (ICISC)*. doi: 10.1109/icisc.2018.8398902.

15. Shwe, H. Y., & Chong, P. H. J. (2016). Scalable distributed cloud data storage service for internet of things. *2016 International IEEE Conferences on Ubiquitous Intelligence & Computing, Advanced and Trusted Computing, Scalable Computing and Communications, Cloud and Big Data Computing, Internet of People, and Smart World Congress (UIC/ATC/ScalCom/CBDCom/IoP/SmartWorld)*. doi: 10.1109/uic-atc-scalcom-cbdcom-iop-smartworld.2016.0137.

16. Ghafour, S. A., Barhamgi, M., & Ghodous, P. (2014). On-demand data integration on the cloud. *2014 IEEE 7th International Conference on Cloud Computing*. doi: 10.1109/cloud.2014.127.

17. Blundo, C., & D'Arco, P. (2005). Analysis and design of distributed key distribution centers. *Journal of Cryptology*, 18(4), 391–414. doi: 10.1007/s00145-005-0407-0.

18. Yang, Y., Lu, H., & Weng, J. (2011). Multi-user private keyword search for cloud computing. *2011 IEEE Third International Conference on Cloud Computing Technology and Science*. doi: 10.1109/cloudcom.2011.43.

19. Liu, C., Zhang, X., Liu, C., Yang, Y., Ranjan, R., Georgakopoulos, D., & Chen, J. (2013). An iterative hierarchical key exchange scheme for secure scheduling of big data applications in cloud computing. *2013 12th IEEE International Conference on Trust, Security and Privacy in Computing and Communications*. doi: 10.1109/trustcom.2013.65.

Index

For Product Safety Concerns and Information please contact our EU
representative GPSR@taylorandfrancis.com
Taylor & Francis Verlag GmbH, Kaufingerstraße 24, 80331 München, Germany

www.ingramcontent.com/pod-product-compliance
Lightning Source LLC
Chambersburg PA
CBHW060343220326
41598CB00023B/2790

9 781032 114262